Level 2

BENCHMARK SERIES

Microsoft® Access®

365

2019 Edition

Jan Davidson
Lambton College
Sarnia, Ontario

PARADIGM
EDUCATION SOLUTIONS

St. Paul

Vice President, Content and Digital Solutions: Christine Hurney
Director of Content Development: Carley Fruzzetti
Developmental Editor: Jennifer Joline Anderson
Director of Production: Timothy W. Larson
Production Editor/Project Manager: Jen Weaverling
Senior Design and Production Specialist: Jack Ross
Cover and Interior Design: Valerie King
Copy Editor: Communicáto, Ltd.
Testers: Janet Blum, Lisa Hart
Indexer: Terry Casey
Vice President, Director of Digital Products: Chuck Bratton
Digital Projects Manager: Tom Modl
Digital Solutions Manager: Gerry Yumul
Senior Director of Digital Products and Onboarding: Christopher Johnson
Supervisor of Digital Products and Onboarding: Ryan Isdahl
Vice President, Marketing: Lara Weber McLellan
Marketing and Communications Manager: Selena Hicks

Care has been taken to verify the accuracy of information presented in this book. However, the authors, editors, and publisher cannot accept responsibility for web, email, newsgroup, or chat room subject matter or content, or for consequences from the application of the information in this book, and make no warranty, expressed or implied, with respect to its content.

Trademarks: Microsoft is a trademark or registered trademark of Microsoft Corporation in the United States and/or other countries. Some of the product names and company names included in this book have been used for identification purposes only and may be trademarks or registered trade names of their respective manufacturers and sellers. The authors, editors, and publisher disclaim any affiliation, association, or connection with, or sponsorship or endorsement by, such owners.

Paradigm Education Solutions is independent from Microsoft Corporation and not affiliated with Microsoft in any manner.

Cover Photo Credit: © lowball-jack/GettyImages
Interior Photo Credits: Follow the Index.

We have made every effort to trace the ownership of all copyrighted material and to secure permission from copyright holders. In the event of any question arising as to the use of any material, we will be pleased to make the necessary corrections in future printings.

ISBN 978-0-76388-732-2 (print)
ISBN 978-0-76388-710-0 (digital)

© 2020 by Paradigm Publishing, LLC
875 Montreal Way
St. Paul, MN 55102
Email: CustomerService@ParadigmEducation.com
Website: ParadigmEducation.com

Printed in the United States of America

28 27 26 25 24 23 22 21 20 19   1 2 3 4 5 6 7 8 9 10 11 12

Brief Contents

Contents

Achieving Proficiency in Access

The Benchmark Series, *Microsoft® Access® 365*, 2019 Edition, is designed for students who want to learn how to use Microsoft's feature-rich data management tool to track, report, and share information. After successfully completing a course in Microsoft Access using this courseware, students can expect to be proficient in using Access to do the following:

- Create database tables to organize business or personal records.
- Modify and manage tables to ensure that data is accurate and up to date.
- Perform queries to assist with decision making.
- Plan, research, create, revise, and publish database information to meet specific communication needs.
- Examine a workplace scenario requiring the reporting and analysis of data, assess the information requirements, and then prepare the materials that achieve the goal efficiently and effectively.

Well-designed pedagogy is important, but students learn technology skills through practice and problem solving. Technology provides opportunities for interactive learning as well as excellent ways to quickly and accurately assess student performance. To this end, this course is supported with Cirrus, Paradigm's cloud-based training and assessment learning management system. Details about Cirrus as well as its integrated student courseware and instructor resources can be found on page xii.

Proven Instructional Design

The Benchmark Series has long served as a standard of excellence in software instruction. Elements of the series function individually and collectively to create an inviting, comprehensive learning environment that leads to full proficiency in computer applications. The following visual tour highlights the structure and features that comprise the highly popular Benchmark model.

Microsoft

Access Level 2

Unit 1

Advanced Tables, Relationships, Queries, and Forms

Chapter 1   Designing the Structure of Tables

Chapter 2   Building Relationships and Lookup Fields

Chapter 3   Advanced Query Techniques

Chapter 4   Creating and Using Custom Forms

1

Unit Openers display the unit's four chapter titles. Each level of the course contains two units with four chapters each.

Chapter Openers Present Learning Objectives

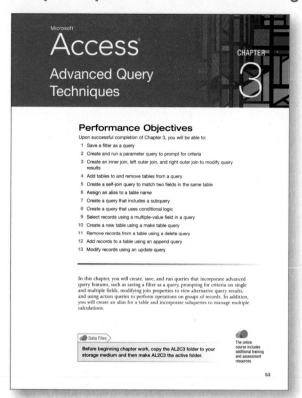

Chapter Openers present the performance objectives and an overview of the skills taught.

Data Files are provided for each chapter.

Activities Build Skill Mastery within Realistic Context

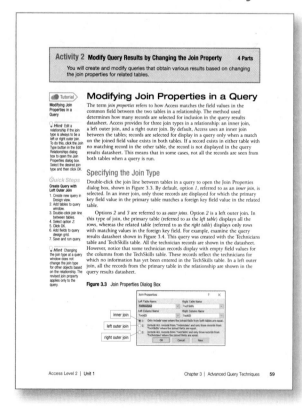

Multipart Activities provide a framework for instruction and practice on software features. An activity overview identifies tasks to accomplish and key features to use in completing the work.

Typically, a file remains open throughout all parts of the activity. Students save their work incrementally. At the end of the activity, students save, print, and then close the file.

Tutorials provide interactive, guided training and measured practice.

Quick Steps in the margins allow fast reference and review.

Hints offer useful tips on how to use features efficiently and effectively.

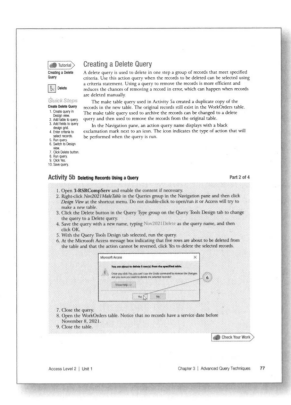

Between activity parts, the text presents instruction on the features and skills necessary to accomplish the next portion of the activity.

Step-by-Step Instructions guide students to the desired outcome for each project part. Screen captures illustrate what the screen should look like at key points.

Magenta Text identifies material to type.

Check Your Work model answer images are available in the online course, and students can use those images to confirm they have completed the activity correctly.

Chapter Review Tools Reinforce Learning

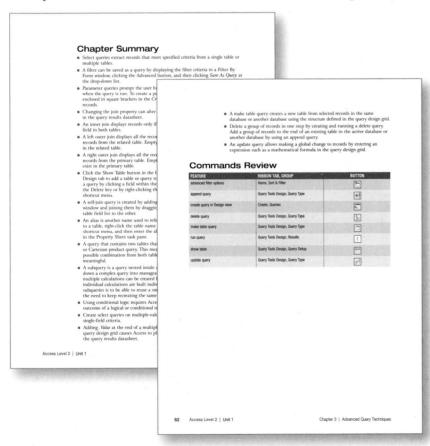

A **Chapter Summary** reviews the purpose and execution of key features.

A **Commands Review** summarizes visually the major software features and alternative methods of access.

The Cirrus Solution
Elevating student success and instructor efficiency

Powered by Paradigm, Cirrus is the next-generation learning solution for developing skills in Microsoft Office. Cirrus seamlessly delivers complete course content in a cloud-based learning environment that puts students on the fast track to success. Students can access their content from any device anywhere, through a live internet connection; plus, Cirrus is platform independent, ensuring that students get the same learning experience whether they are using PCs, Macs, or Chromebook computers.

Cirrus provides Benchmark Series content in a series of scheduled assignments that report to a grade book to track student progress and achievement. Assignments are grouped in modules, providing many options for customizing instruction.

Dynamic Training

The online Benchmark Series courses include interactive resources to support learning.

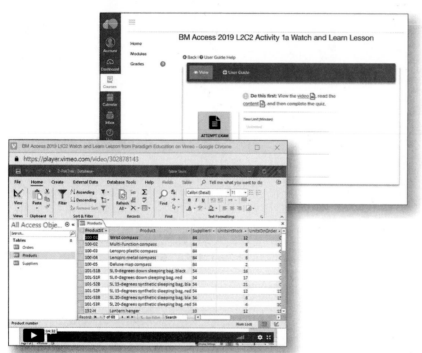

Watch and Learn Lessons include a video demonstrating how to perform the chapter activity, a reading to provide background and context, and a short quiz to check understanding of concepts and skills.

Guide and Practice Tutorials provide interactive, guided training and measured practice.

Hands On Activities enable students to complete chapter activities, compare their solutions against a Check Your Work model answer image, and submit their work for instructor review.

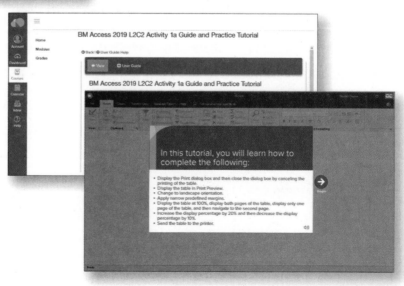

Chapter Review and Assessment

Review and assessment activities for each chapter are available for completion in Cirrus.

Concepts Check completion exercises assess their comprehension and recall of application features and functions as well as key terminology.

Skills Assessment Hands On Activity exercises evaluate the ability to apply chapter skills and concepts in solving realistic problems. Each is completed live in Access and is uploaded through Cirrus for instructor evaluation.

Visual Benchmark assessments test problem-solving skills and mastery of application features.

A **Case Study** requires analyzing a workplace scenario and then planning and executing a multipart project. Students search the web and/or use the program's Help feature to locate additional information required to complete the Case Study.

Exercises and **Projects** provide opportunities to develop and demonstrate skills learned in each chapter. Each is completed live in the Office application and is automatically scored by Cirrus. Detailed feedback and how-to videos help students evaluate and improve their performance.

Skills Check Exams evaluate students' ability to complete specific tasks. Skills Check Exams are completed live in the Office application and are scored automatically. Detailed feedback and instructor-controlled how-to videos help student evaluate and improve their performance.

Multiple-choice **Concepts Exams** assess understanding of key commands and concepts presented in each chapter.

Unit Review and Assessment

Review and assessment activities for each unit of each Benchmark course are also available for completion in Cirrus.

Assessing Proficiency exercises check mastery of software application functions and features.

Writing Activities challenge students to use written communication skills while demonstrating their understanding of important software features and functions.

Internet Research assignments reinforce the importance of research and information processing skills along with proficiency in the Office environment.

A **Job Study** activity at the end of Unit 2 presents a capstone assessment requiring critical thinking and problem solving.

Unit-Level Projects allow students to practice skills learned in the unit. Each is completed live in the Office application and automatically scored by Cirrus. Detailed feedback and how-to videos help students evaluate and improve their performance.

Student eBook

The Student eBook, accessed through the Cirrus online course, can be downloaded to any device (desktop, laptop, tablet, or smartphone) to make Benchmark Series content available anywhere students wish to study.

Instructor eResources

Cirrus tracks students' step-by-step interactions as they move through each activity, giving instructors visibility into their progress and missteps. With Exam Watch, instructors can observe students in a virtual, live, skills-based exam and join remotely as needed—a helpful option for struggling students who need one-to-one coaching, or for distance learners. In addition to these Cirrus-specific tools, the Instructor eResources for the Benchmark Series include the following support:

- Planning resources, such as lesson plans, teaching hints, and sample course syllabi
- Delivery resources, such as discussion questions and online images and templates
- Assessment resources, including live and annotated PDF model answers for chapter work and review and assessment activities, rubrics for evaluating student work, and chapter-based exam banks in RTF format

About the Author

Jan Davidson started her teaching career in 1997 as a corporate trainer and postsecondary instructor and holds a Social Science degree, a writing certificate, and an In-Service Teacher Training certificate. Since 2001, she has been a faculty member of the School of Business and International Education at Lambton College in Sarnia, Ontario. In this role, she has developed curriculum and taught a variety of office technology, software applications, and office administration courses to domestic and international students in a variety of postsecondary programs. As a consultant and content provider for Paradigm Education Solutions since 2006, Jan has contributed to textbook and online content for various titles. She has been author and co-author of Paradigm's Benchmark Series *Microsoft® Excel®*, Level 2, and *Microsoft® Access®*, Level 2 since 2013 and has contributed to the Cirrus online courseware for the series. Jan is also co-author of *Advanced Excel® 2016*.

Microsoft Access® Level 2

Unit 1

Advanced Tables, Relationships, Queries, and Forms

Microsoft®
Access®

Designing the Structure of Tables

Performance Objectives

Upon successful completion of Chapter 1, you will be able to:

1 Design the structure of a table

2 Select field data type based on an analysis of the source data

3 Disallow blank field values

4 Allow and disallow zero-length strings in fields

5 Create custom formats for Short Text, Number, and Date/Time data type fields

6 Restrict data entry using input masks

7 Enable rich text formatting in a Long Text data type field

8 Maintain a history of changes for a Long Text data type field

9 Define and use an Attachment data type field with multiple attachments

---

Designing tables in Access is the most important task when creating a database because all the other objects are based on tables. All queries, forms, and reports rely on tables as the sources of their data. Designing a new database involves planning the number of tables needed, the fields to be included in each table, and the methods to be used to check and/or validate new data as it is entered. In this chapter, you will learn the basic steps involved in planning a new database by analyzing existing data. You will also learn to select appropriate data types and use field properties to control, restrict, or otherwise validate data.

In order to complete the activities in this chapter, you will already need to know the steps in creating a new table, including changing the data type and field size and assigning the primary key field. Readers are also assumed to know the meanings of the terms *field*, *record*, *table*, and *database*.

 Data Files

Before beginning chapter work, copy the AL2C1 folder to your storage medium and then make AL2C1 the active folder.

The online course includes additional training and assessment resources.

Tutorial

Review: Creating
a Table in Design
View

Designing Tables and
Fields for a New Database

Most databases encountered in the workplace have been created by database
designers. Even so, understanding the process involved in creating a new database
will help users make sense of how objects are organized and related. Creating a new
database from scratch involves careful planning.

Database designers spend considerable time analyzing existing data and asking
questions of users and managers. Designers want to know how data will be used
so they can identify the forms, queries, and reports that will need to be generated.
Often designers begin by modeling a required report from the database to see the
data used to populate the report. The designer then compiles a data dictionary,
which is a list of fields as well as the attributes of each field. The designer uses the
data dictionary to map out the number of required tables.

A sample work order for RSR Computer Services is analyzed in Activity 1.
RSR started as a small computer service company and the owners used Excel
worksheets to enter information from service records and to produce revenue
reports. The company's success has created the need for a relational database to
track customer information. The owners want to be able to generate queries and
reports that will help them in decision making. Examine the data in the sample
work order shown in Figure 1.1. The work order that the technicians have been
filling out at the customer site will be used as the input source document for the
database.

Figure 1.1 Sample Work Order for RSR Computer Services

Designers analyze all the input documents and output requirements to capture the entire set of data elements that needs to be created. Once all the data has been identified, the designer maps out the number of tables required to hold it. During the process of mapping out the tables and fields to be associated with each table, the designer follows these guidelines and techniques:

- Consider each table an entity that describes a single person, place, object, event, or other subject. Each table should store facts related only to that entity.

- Segment the data until it is in its smallest unit. For example, in the work order shown in Figure 1.1, the customer's name and address should be split into separate fields for first name, last name, street address, city, state, and zip code. Using this approach provides maximum flexibility for generating other objects and allows the user to sort or filter by any individual data element.

- Do not include fields that can be calculated using data from other fields. For example, the total labor and total due amounts in the work order can be calculated using other elements of numeric data.

- Identify fields that can be used to answer questions from the data. Queries and reports can be designed to extract information based on the results of a conditional expression (sometimes referred to as Boolean logic). For example, in the work order in Figure 1.1, the technician indicates whether the customer has a service contract. Providing a field that stores a *Yes* or *No* (true or false) condition for the service contract data element allows the business to generate reports of customers that have subscribed to a service contract (true condition) and customers that have not subscribed to a service contract (false condition).

- Identify a field in each table that will hold the data that uniquely identifies each record. This field becomes the primary key field. If the data that the database is designed to organize do not reveal a logical unique identifier, it is possible to use the ID field that Access automatically generates with the AutoNumber data type for each record as the primary key field.

- Identify each table that will relate to another table and the field that will be used to join the two tables when the relationships are created. Identifying relationships at this stage helps determine whether a field needs to be added to a related table to allow the tables to be joined.

- Keep in mind that relational databases are built on the concept that data redundancy should be avoided, except for fields that will be used to join tables in a relationship. (The term *data redundancy* means that data in one table is repeated in another table.) Repeating fields in multiple tables wastes storage space, promotes inefficiency and inconsistency, and increases the likelihood of errors being made when adding, updating, and deleting field values.

The database design process may seem time consuming, but creating a well-designed database will save time later. A poorly designed database often contains logical and structural errors that require redefining data or objects after live data has been entered.

Diagramming a Database

Recall from Level 1, Chapter 1 that designers often create a visual representation of the structure of a database in a diagram similar to the one shown in Figure 1.2. In the database diagram, each table is represented in a box with the table name at the top. Within each box, the fields that will be stored in the table are listed with the field names that will be used when the table is created. The

Figure 1.2 Diagram of Table Structure for RSR Computer Services Database

Customers
*CustID
FName
LName
Street
City
State
ZIP
HPhone
CPhone
ServCont

ServiceContracts
*CustID
SCNo
StartDate
EndDate
FeePd

Technicians
*TechID
SSN
FName
LName
Street
City
State
ZIP
HPhone
CPhone

WorkOrders
*WO
CustID
TechID
WODate
Descr
ServDate
Hours
Rate
Parts
Comments

Hint Words such as *Name* and *Date* are reserved words in Access and cannot be used as field names. A prompt will appear when trying to save a table containing field names that use reserved words.

primary key field is denoted with an asterisk. Tables that will be joined are connected with lines at their common fields. The database represented in Figure 1.2 will be built in the remainder of this chapter and the relationships will be created in Chapter 2.

Notice that many of the field names in the diagram are abbreviated. Although a field name can contain up to 64 characters, using field names that are short enough to be understood is recommended; they are easier to manage and to type into expressions. For abbreviated field names, the Caption property is used to display descriptive headings that contain spaces and/or longer words when viewing the data in a datasheet, form, or report.

Also notice that none of the field names contains spaces. Spaces are allowed in field names but most database designers avoid using them. Instead, designers indicate a space between words by changing the case, using an underscore character (_), or using a hyphen (-).

Assigning Data Types

Designers assign each field a data type based on the types of entries that will be allowed into the field and the operations that will be used to manipulate the data. Selecting the appropriate data type is important because restrictions will be placed on a field based on its data type. For example, in a field designated with the Number data type, only numbers, a period to represent a decimal point, and a plus or minus symbol (+ or –) can be entered into the field in a datasheet or form. Table 1.1 identifies the available data types.

Table 1.1 Data Types

Data Type	Description
Short Text	Alphanumeric data up to 255 characters—for example, a name, an address, or a value such as a telephone number or social security number that is used as an identifier and not for calculating.
Long Text	Alphanumeric data longer than 255 characters; up to 65,535 characters can be stored in a field, although only 64,000 can be displayed. These fields are used to store longer passages of text in a record. Rich text formatting can be added such as bold, italic, or font color.
Number	Positive or negative values that can be used in calculations; not to be used for monetary amounts (see Currency).
Large Number	Non-monetary, numeric values used to calculate large numbers. Not backwards compatible with previous Access versions.
Date/Time	Used to ensure dates and times are entered and sorted properly.
Currency	Values that involve money; Access will not round off during calculations.
AutoNumber	Used to automatically number records sequentially (increments of 1); each new record is numbered as it is typed.
Yes/No	Data in the field is restricted to conditional logic of *Yes* or *No*, *True* or *False*, *On* or *Off*.
OLE Object	Used to embed or link objects in other Office applications.
Hyperlink	Used to store a hyperlink, such as a URL.
Attachment	Used to add file attachments to a record such as a Word document or Excel workbook.
Calculated	Used to display the Expression Builder dialog box, where an expression is entered to calculate the value of the calculated column.
Lookup Wizard	Used to enter data in a field from another existing table or to display a list of values in a drop-down list from which the user chooses.

Using the Field Size Property to Restrict Field Length

By default, a Short Text data type field is set to a width of 255 characters in the Field Size property. Access uses only the amount of space needed for the data entered, even when the field size allows for more characters. Even so, it can be helpful to change this property to a smaller value.

One reason to change the Field Size property to a smaller value is that it restricts the length of the data that can be entered into the field. For example, assume that RSR Computer Services has developed a four-character numbering system for customer numbers. Setting the field size for the *CustID* field to four characters will ensure that no one enters a longer customer number by accident. Access will disallow any character typed after the fourth character.

Figure 1.3 shows the table structure diagram for the RSR Computer Services database expanded to include the data type and Field Size property for each field. Use this diagram to create the tables in Activity 1a.

Figure 1.3 Expanded Table Structure Diagram with Data Types and Field Sizes for Activity 1a

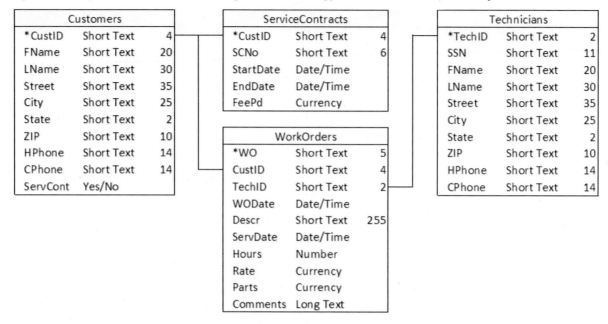

Customers		
*CustID	Short Text	4
FName	Short Text	20
LName	Short Text	30
Street	Short Text	35
City	Short Text	25
State	Short Text	2
ZIP	Short Text	10
HPhone	Short Text	14
CPhone	Short Text	14
ServCont	Yes/No	

ServiceContracts		
*CustID	Short Text	4
SCNo	Short Text	6
StartDate	Date/Time	
EndDate	Date/Time	
FeePd	Currency	

WorkOrders		
*WO	Short Text	5
CustID	Short Text	4
TechID	Short Text	2
WODate	Date/Time	
Descr	Short Text	255
ServDate	Date/Time	
Hours	Number	
Rate	Currency	
Parts	Currency	
Comments	Long Text	

Technicians		
*TechID	Short Text	2
SSN	Short Text	11
FName	Short Text	20
LName	Short Text	30
Street	Short Text	35
City	Short Text	25
State	Short Text	2
ZIP	Short Text	10
HPhone	Short Text	14
CPhone	Short Text	14

Activity 1a Creating Tables in Design View

Part 1 of 6

1. Start Access.
2. At the Access 365 opening screen, complete the following steps to create a new database to store the work orders for RSR Computer Services:
   a. Click the *Blank database* template.
   b. At the Blank database window, click the Browse button and navigate to the AL2C1 folder on your storage medium. Select the current text in the *File Name* text box, type 1-RSRCompServ, and then click OK.
   c. Click the Create button below the *File Name* text box.
3. Close the Table1 blank table datasheet that displays. Design view will be used to access all the field properties available for fields.
4. Click the Create tab and then click the Table Design button in the Tables group.
5. Create the fields shown in the Customers table in Figure 1.3, including the data type and field size settings, by completing the following steps:
   a. Type CustID and then press the Tab key. Short Text is the default data type.
   b. Double-click *255* in the *Field Size* property box in the *Field Properties* section and then type 4.
   c. Click in the blank *Field Name* text box directly under the *CustID* field.
   d. Enter the rest of the field names, data types, and field sizes for the Customers table shown in Figure 1.3 by completing steps similar to those in Steps 5a–c. For the *ServCont* data type, type y or click the *Data Type* option box arrow and then select *Yes/No*.

6. Select the *CustID* field and then click the Primary Key button in the Tools group.
7. Click the Save button on the Quick Access Toolbar, type Customers in the *Table Name* text box, and then click OK.

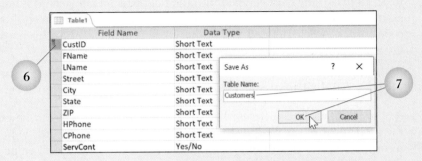

8. Close the table.
9. Create the ServiceContracts, WorkOrders, and Technicians tables shown in Figure 1.3 by completing steps similar to those in Steps 4–8. Assign the primary key field in each table using the fields denoted with asterisks in Figure 1.3 (on page 8).
10. Make sure all the tables are closed.

Controlling Data Display and Data Entry Using Field Properties

Tutorial

Controlling Data Entry

Controlling Data Display and Data Entry Using Field Properties

What properties are available for a field depend on the field's data type. For example, a Yes/No data type field has 7 properties, whereas a Short Text data type field has 14 and a Number data type field has 11.

Use the options available in the *Field Properties* section in Design view to place restrictions on or control data accepted into a field and ensure that data is entered and displayed consistently. Field properties carry over when objects such as queries, forms, and reports are created if the properties are defined before the other objects are created. Taking the time to define the properties when the table is created reduces the number of changes needed if the properties are modified later.

When entering field properties, it is important to select the appropriate field and enter the properties correctly. If properties are applied to the wrong field or entered incorrectly, Access may not be able to save the table or enter the data.

Adding Captions

In Level 1, captions were entered using the Caption property in the Name & Caption dialog box when a new table was created using Datasheet view. The same property appears in Design view in the *Field Properties* section. Recall that the Caption property allows the user to enter a more descriptive title for a field if the field name has been truncated or abbreviated.

The words in field name captions should be separated using spaces, rather than the underscore or hyphen characters used in the field names themselves. In the absence of a caption, Access displays the field name in datasheets, queries, forms, and reports.

Quick Steps

Add Caption to Existing Field
1. Open table in Design view.
2. Select field.
3. Click in *Caption* property box.
4. Type descriptive text.
5. Save table.

Require Data in Field
1. Open table in Design view.
2. Select field.
3. Double-click in *Required* property box.
4. Save table.

Requiring Data in a Field

Hint Set the *Required* field to *Yes* and *Allow Zero Length* to *No* to make sure a field value (and not a space) will be entered when the record is added.

Use the Required property to make sure that a certain field is never left empty when a new record is added. By default, the Required property is set to *No*. Change this value to *Yes* to make sure data is typed into the field when a new record is added. For example, setting the Required property to *Yes* will force all new records to have zip code entries. A field defined as a primary key field already has this property set to *Yes* because a primary key field cannot be left empty.

Using and Disallowing Zero-Length Strings

Quick Steps

Disallow Zero-Length String in Field
1. Open table in Design view.
2. Activate a field.
3. Double-click in *Allow Zero Length* property box.
4. Save table.

Hint Press the spacebar to insert a zero-length string.

A zero-length field can be used to indicate that a value will not be entered into the field because the field does not apply to the current record. When a new record is entered and a field is left blank, Access records a null value in the field. For example, assume that a new record for a customer is being entered and the cell phone number is not known; leave the field empty with the intention of updating it later. Access will record a null value in the field. Alternatively, assume that if there is no home phone number; enter a zero-length string in the field to indicate that no field value applies to this record.

To enter a zero-length string, type two double quotation marks with no space between them (""). It is impossible to distinguish between a field with a null value and a field with a zero-length string when viewing the field in a datasheet, query, form, or report; both display as blanks. To help distinguish between the two, create a control in a form or report that returns a user-defined message in a blank field. For example, display the word *Unknown* in a field with a null value and the phrase *Not applicable* in a field with a zero-length string.

By default, Short Text, Long Text, and Hyperlink data type fields allow zero-length strings. Change the Allow Zero Length property to *No* to disallow zero-length strings.

Activity 1b Modifying Field Properties to Add Captions and Disallow Blank Values in a Field

Part 2 of 6

1. With **1-RSRCompServ** open, add captions to the fields in the Customers table by completing the following steps:
   a. Right-click *Customers* in the Tables group in the Navigation pane and then click *Design View* at the shortcut menu.
   b. With *CustID* the active field, click in the *Caption* property box in the *Field Properties* section and then type Customer ID.
   c. Click in the *FName* field row to activate the field, click in the *Caption* property box in the *Field Properties* section, and then type First Name.
   d. Add captions to the following fields by completing the step similar to Step 1c:

LName	Last Name
HPhone	Home Phone
CPhone	Cell Phone
ServCont	Service Contract?

e. Click the Save button on the Quick Access Toolbar.

f. Click the View button (not the button arrow) to switch to Datasheet view and then select all the columns in the datasheet. If necessary, click the Shutter Bar Open/Close button (the two left-pointing chevrons at the top of the Navigation pane) to minimize the Navigation pane.

g. Click the More button in the Records group on the Home tab, click *Field Width* at the drop-down list, and then click the Best Fit button at the Column Width dialog box to adjust the column widths to fit the longest entries.

h. To deselect the columns, click in the *Customer ID* field in the first row of the datasheet.

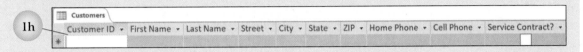

2. Switch to Design View and then click the Shutter Bar Open/Close button (the two right-pointing chevrons) to redisplay the Navigation pane if the pane was minimized in Step 1f.

3. Ensure that no record is entered without an entry in the *ZIP* field and disallow blank values in the field, including zero-length strings, by completing the following steps:

a. Click in the *ZIP* field row to select the field.

b. Click in the *Required* property box in the *Field Properties* section (which displays *No*), click the option box arrow that appears, and then click *Yes* at the drop-down list.

c. Double-click in the *Allow Zero Length* property box (which displays *Yes*) to change the *Yes* to *No*.

d. Save the changes to the table design.

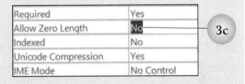

4. Use a new record to test the restrictions on the *ZIP* field by completing the following steps:

a. Switch to Datasheet view.

b. Add the following data in the fields indicated:

Customer ID	1000
First Name	Jade
Last Name	Fleming
Street	12109 Woodward Avenue
City	Detroit
State	MI

c. At the *ZIP* field, press the Tab key to move past the field, leaving it blank.

d. Type 313-555-0214 in the *Home Phone* field.

e. Type 313-555-3485 in the *Cell Phone* field.

f. Press the spacebar in the *Service Contract?* field to insert a check mark in the check box.

g. Press the Enter key. Access displays an error message because the record cannot be saved without an entry in the *ZIP* field.

h. Click OK at the Microsoft Access message box.

i. Click in the *ZIP* field, type 48203-3579, and then press the Enter key four times to move to the *Customer ID* field in the second row of the datasheet.

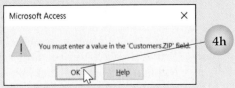

5. Double-click the right boundaries of the *Street* and *ZIP* columns to adjust the column widths so that the entire field values can be read.

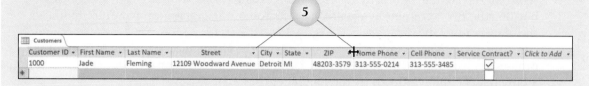

6. Close the Customers table. Click Yes when prompted to save changes to the layout of the table.

Check Your Work

Tutorial

Creating a Custom
Format for a Short
Text Field

Quick Steps

**Format Short Text
Field**
1. Open table in Design
   view.
2. Activate field.
3. Click in *Format*
   property box.
4. Type format codes.
5. Save table.

Creating a Custom Format for a Short Text Data Type Field

The Format property controls how data is displayed in the field in the datasheet, query, form, or report. What formats are available depends on the data type of the field. Predefined formats are available for some data types and can be selected from a drop-down list in the *Format* property box. No predefined formats exist for Short Text and Long Text data type fields. If no predefined format exists or the predefined format options are not suitable, then a custom format can be created. Table 1.2 displays commonly used format codes for Short Text and Long Text data type fields.

The Format property does not control how data is entered into the field. Rather, formatting a field controls the display of accepted field values. Refer to the section on input masks (which begins on page 17) to learn how to control new data as it is being entered.

Table 1.2 Format Codes for Short Text and Long Text Data Type Fields

Code	Description	Format Property Example
@	Used as a placeholder, one symbol for each character position. An unused position in a field value is replaced with a blank space to the left of the text entered into the field.	@@@@ Field value entered is *123*. Access displays one blank space followed by *123*, left-aligned in the field.
!	Placeholder positions are filled with characters from left to right instead of the default right to left sequence.	!@@@@ Field value entered is *123*. Access displays *123* left-aligned in the field with one blank space after the *3*.
>	All text is converted to uppercase.	> Field value entered is *mi*. Access displays *MI* in the field.
<	All text is converted to lowercase.	< Field value entered is *Jones@EMCP.NET*. Access displays *jones@emcp.net* in the field.
[color]	Text is displayed in the font color specified. Available colors are black, blue, cyan, green, magenta, red, yellow, and white.	[red]@@@@@-@@@@ Field value entered is *482033579*. Access displays *48203-3579*.

1. With **1-RSRCompServ** open, format the *State* field in the Customers table to ensure that all the text is displayed in uppercase letters by completing the following steps:
   a. Right-click *Customers* in the Tables group in the Navigation pane and then click *Design View* at the shortcut menu.
   b. Click in the *State* field row to activate the field.
   c. Click in the *Format* property box and then type >.
   d. Save the table.

General	Lookup		
Field Size		2	
Format		>	1c
Input Mask			

2. Format the *ZIP* field to fill it with characters from left to right, display the text in red, and provide for the five-plus-four-character US zip code (with the two sets of characters, each separated by a hyphen) by completing the following steps:
   a. Click in the *ZIP* field row to activate the field.
   b. Click in the *Format* property box and then type ![red]@@@@@-@@@@.
   c. Save the table.

General	Lookup		
Field Size		10	
Format		![red]@@@@@-@@@@	2b
Input Mask			

3. Test the custom formats in the *State* and *ZIP* fields using a new record by completing the following steps:
   a. Switch to Datasheet view.
   b. Add the following data in a new record. Type the text for the *State* field as indicated in lowercase text. Notice that Access automatically converts the lowercase text to uppercase when moving to the next field. As you type the *ZIP* field text, notice that it displays in red. Since no field values are entered for the last four characters of the *ZIP* field, Access displays blank spaces in these positions.

Customer ID	1005
First Name	Cayla
Last Name	Fahri
Street	12793 Riverdale Avenue
City	Detroit
State	mi
ZIP	48223
Home Phone	313-555-6845
Cell Phone	313-555-4187
Service Contract?	Press the spacebar for *Yes*.

4. Look at the data in the *ZIP* field for the first record. This data was entered before the *ZIP* field was formatted. Since a hyphen was typed when the data was entered in Activity 1b and the field is now formatted to automatically add the hyphen, two hyphen characters appear in the existing record. Update the record by editing the *ZIP* field value for record 1 to remove the extra hyphen.

5. Display the datasheet in Print Preview. Change to landscape orientation. Set the top margin to 1 inch and the bottom, left, and right margins to 0.25 inch. Print the datasheet and then close Print Preview.
6. Close the Customers table.

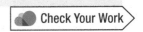
Check Your Work

Ȯuick Steps
Format Numeric Field
1. Open table in Design
   view.
2. Activate field.
3. Click in *Format*
   property box.
4. Type format codes
   or select from
   predefined list.
5. Save table.

Creating a Custom Format for a Numeric Data Type Field

Access provides predefined formats for Number, AutoNumber, and Currency data type fields that include options for displaying a fixed number of digits past the decimal point, a comma separator, and a currency symbol. Table 1.3 displays format codes that can be used to create custom formats. Use the placeholders shown in Table 1.3 in combination with other characters (such as the dollar symbol, comma, and period) to create the desired custom numeric format.

Specify up to four formats for a numeric data type field to include different options for displaying positive values, negative values, zero (0), and null values. Examine the following custom format code:

#,###.00;-#,###.00[red];0.00;"Unknown"

Notice that the four sections are separated with semicolons (;). The first section, *#,###.00*, defines the format for positive values. It includes a comma in thousands values and two digits past the decimal point; zeros are used if no decimal value is entered. The second section, *-#,###.00[red]*, defines negative values. It includes the same placeholders as used with positive values but starts the field with a minus symbol and displays the numbers in red. The color code can be placed before or after the codes for the numbers. The third section, *0.00*, instructs Access to show *0.00* in the field if a zero is entered. Finally, a field value that is a null value (the spacebar is pressed) will display the text *Unknown* (italic used here for emphasis only) in the field. Text to be directly entered into the field is enclosed in quotation marks.

Table 1.3 Format Codes for Numeric Data Type Fields

Code	Description	Format Property Example
#	Used as a placeholder to display a number.	*#.##* Field value entered is *123.45*. Access displays *123.45* in the field. Notice that the number of placeholder positions does not restrict the data entered into the field.
0	Used as a placeholder to display a number. Access displays a *0* in place of a position for which no value is entered.	*000.00* Field value entered is *55.4*. Access displays *055.40* in the field.
%	Value is multiplied times 100 and a percent symbol is added.	*#.0%* Field value entered is *.1246*. Access displays *12.5%* in the field. Field value entered is *.1242*. Access displays *12.4%* in the field. Having only one digit past the decimal point causes rounding up or down to occur.

1. With **1-RSRCompServ** open, format the *Rate* field in the WorkOrders table with a custom format that displays positive numbers with two decimal places and blue text and null values with the text *Not Available* by completing the following steps:
   a. Open the WorkOrders table in Design view.
   b. Make the *Rate* field active.
   c. Click in the *Format* property box, delete the current entry (*Currency*), and then type #.00[blue];;;"Not Available". Notice that three semicolons are typed after the first custom format option, *#.00[blue]*. When a custom format for negative or zero values is not needed, include the semicolon to indicate that there is no format setting. Since an hourly rate is never a negative or zero value, do not use custom formats in these situations.
   d. Save the table.

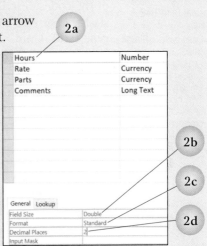

2. Change the *Hours* field size to one that displays fractional numbers and format the field using a predefined format by completing the following steps:
   a. Make the *Hours* field active.
   b. Click in the *Field Size* property box, click the option box arrow that appears, and then click *Double* at the drop-down list. The default setting for a Number data type field is *Long Integer*, which stores only whole numbers. (This means that a decimal value entered into the field is rounded.) Changing the *Field Size* property box to *Double* allows decimal values to be stored.
   c. Click in the *Format* property box, click the option box arrow that appears, and then click *Standard* at the drop-down list.
   d. Click in the *Decimal Places* property box, click the option box arrow that appears, and then click *2* at the pop-up list.
   e. Save the table.
3. Switch to Datasheet view.
4. Add the following data in a new record to test the custom format and the predefined format. Notice that when you move past the *Rate* field, the value is displayed in blue. Compare your data entry to the screen image on the next page and notice the new format for the *Hours* and *Rate* fields.

WO	65012
CustID	1000
TechID	11
WODate	09-07-2021
Descr	Biannual desktop computer cleaning and maintenance
ServDate	09-07-2021
Hours	1.25
Rate	50
Parts	35.15
Comments	Keys are sticking; cleaning did not resolve. Customer is considering buying a new keyboard.

5. Review the WorkOrders table and note the formatting.

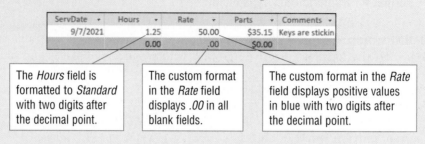

ServDate	Hours	Rate	Parts	Comments
9/7/2021	1.25	50.00	$35.15	Keys are stickin
	0.00	.00	$0.00	

The *Hours* field is formatted to *Standard* with two digits after the decimal point.

The custom format in the *Rate* field displays *.00* in all blank fields.

The custom format in the *Rate* field displays positive values in blue with two digits after the decimal point.

6. Close the WorkOrders table.

Creating a Custom Format for a Date/Time Field

Creating a Custom Format for a Date/Time Field

Access provides predefined formats for fields with a Date/Time data type. These formats provide a variety of combinations of month, day, and year options for dates and hour and minute display options for time. If the predefined display formats are not suitable, custom formats can be created using a combination of the codes described in Table 1.4, along with symbols (such as hyphens and slashes) between parts of the date. If a format option for a Date/Time data type field is not chosen, Access displays the date in the format *m/d/yyyy*. For example, in Activity 1d, the date entered into the *WODate* field displays as *9/7/2021*.

A custom format for a Date/Time data type field can contain two sections separated by a semicolon. The first section specifies the format for displaying dates. To add a format for displaying times, type a semicolon and then add the format codes.

Quick Steps

Format Date/Time Field

1. Open table in Design view.
2. Activate field.
3. Click in *Format* property box.
4. Type format codes or select from predefined list.
5. Save table.

Table 1.4 Format Codes for Date/Time Data Type Fields

Code	Description
d or dd	displays the day of the month as one digit (*d*) or two digits (*dd*)
ddd or dddd	displays the day of the week abbreviated (*ddd*) or in full (*dddd*)
m or mm	displays the month as one digit (*m*) or two digits (*mm*)
mmm or mmmm	displays the month abbreviated (*mmm*) or in full (*mmmm*)
yy or yyyy	displays the year as the last two digits (*yy*) or all four digits (*yyyy*)
h or hh	displays the hour as one digit (*h*) or two digits (*hh*)
n or nn	displays the minutes as one digit (*n*) or two digits (*nn*)
s or ss	displays the seconds as one digit (*s*) or two digits (*ss*)
AM/PM	displays 12-hour clock values followed by *AM* or *PM*

1. With **1-RSRCompServ** open, format the *WODate* field in the WorkOrders table with a custom format by completing the following steps:
   a. Open the WorkOrders table in Design view.
   b. Make *WODate* the active field.
   c. Click in the *Format* property box, type ddd, mmm dd yyyy, and then press the Enter key. This format will display dates beginning with the day of the week in abbreviated form, followed by a comma, the month in abbreviated form, the day of the month as two digits, and the year as four digits. Spaces separate the sections of the date. Notice that Access puts quotation marks around the comma.

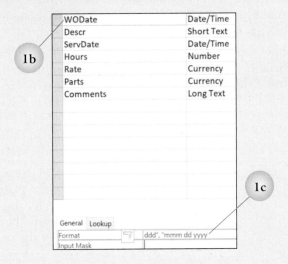

   d. Save the table.
2. Switch to Datasheet view.
3. If necessary, adjust the column width of the *WODate* field to allow reading the entire entry.

Custom format for *WODate* field created in Step 1c.

4. Switch to Design view.
5. Format the *ServDate* field using the same custom format used for *WODate* by completing steps similar to those in Steps 1b–1c.
6. Save the table and then switch to Datasheet view.
7. Double-click the right column boundary to adjust the column width of the *ServDate* field and view the custom date format.

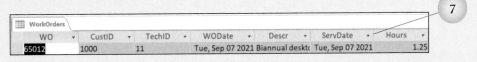

8. Close the WorkOrders table. Click Yes when prompted to save changes to the table layout.

Check Your Work

Quick Steps

Create Custom Input Mask
1. Open table in Design view.
2. Activate field.
3. Click in *Input Mask* property box.
4. Type input mask codes.
5. Save table.

Hint If you create a custom input mask for a date field that also contains a custom format, make sure the two properties do not conflict. For example, a format code that displays dates with the year first followed by the month and then the day will be confusing if the input mask requires the date to be entered as the month first followed by the day and then the year.

Controlling Data Entry Using an Input Mask

Whereas a format controls how the data is displayed, an input mask controls the type of data and pattern in which the data is entered into a field. In Activity 1e, a date format was created to display the date as *Tue, Sep 7 2021*. When confirming dates with clients, it is good practice to provide the day of the week, as well as the day of the month. To type the date consistently and efficiently into the *ServDate* or *WODate* field, an input mask will be created in Activity 1f to enter the date by typing *sep072021*. Access knows that this date is a Tuesday and displays *Tue, Sep 7 2021*.

Recall from Level 1, Chapter 4, that Access provides the Input Mask Wizard, which can be used to create an input mask for a text or date field. Commonly used input masks are predefined within the wizard for telephone numbers, social security numbers, zip codes, dates, and times. To create an input mask without the wizard, use the codes described in Table 1.5.

An input mask can contain up to three sections, with each section separated by a semicolon. The first section contains the input mask codes for the data entry in the field. The second section instructs Access to store or not store the display characters used in the field, such as hyphens, slashes, or brackets). A zero indicates that Access should store the characters. Leaving the second section blank means the display characters will not be stored. The third section specifies the placeholder character to display in the field when it becomes active for data entry.

The following is an example of an input mask to store a four-digit customer identification number with a pound symbol (#) as the placeholder: *0000;;#*. The first section, *0000*, contains the four required digits for the customer identification number. Since the mask contains no display characters (hyphens, slashes, etc.), the second section is blank. The pound symbol after the second semicolon is the placeholder character.

In addition to the symbols in Table 1.5, use the format code > to force characters to be uppercase or the format code < to force characters to be lowercase. Decimal points, hyphens, slashes, and other punctuation symbols can also be used.

Table 1.5 Commonly Used Input Mask Codes

Code	Description
0	Required digit.
9	Optional digit.
#	Digit, space, plus or minus symbol. If no data is typed at this position, Access leaves a blank space.
L	Required letter.
?	Optional letter.
A	Required letter or digit.
a	Optional letter or digit.
&	Required character or space.
C	Optional character or space.
!	Field is filled from left to right instead of right to left.
\	Character is displayed that immediately follows in the field.

1. With **1-RSRCompServ** open, create a custom input mask for the work order numbers in the WorkOrders table by completing the following steps:
   a. Open the WorkOrders table in Design view.
   b. With *WO* the active field, click in the *Input Mask* property box and then type 00000;;_. This mask requires a five-digit work order number to be entered. The underscore is used as the placeholder character that displays when the field becomes active.

   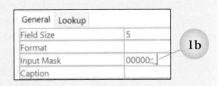

   c. Save the table.
2. Create an input mask to require the two date fields to be entered as three characters for the month, followed by two digits for the day and four digits for the year by completing the following steps:
   a. Make *WODate* the active field.
   b. Click in the *Input Mask* property box and then type >L<LL\ 00\ 0000;0;_. This mask requires three letters for the month; the first letter is converted to uppercase and the remaining two letters are converted to lowercase. The backslash symbol (\ ) followed by the space instructs Access to display a space after the month as data is entered. Two digits are required for the day, followed by another space and then four digits for the year. The *0* after the first semicolon instructs Access to store the display characters. At the end of the mask, the underscore character is used again as the placeholder character.

   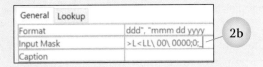

   c. Make *ServDate* the active field, click in the *Input Mask* property box, and then type >L<LL\ 00\ 0000;0;_.
   d. Save the table.
3. Switch to Datasheet view.
4. Test the input masks using a new record by completing the following steps:
   a. Click the New button in the Records group on the Home tab.
   b. Type 6501. Notice that as soon as the first character is typed, the placeholders appear in the field.
   c. Press the Tab key to move to the next field in the datasheet. Since the mask contains five zeros (indicating five required digits), Access displays a message box stating that the value entered is not appropriate for the input mask.
   d. Click OK at the Microsoft Access message box.

   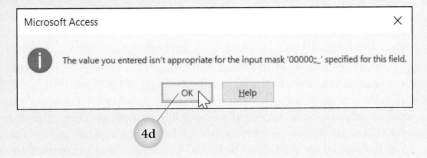

e. Type 3 in the last position in the *WO* field and then press the Tab key to move to the next field.

f. Type 1000 in the *CustID* field and then press the Tab key.

g. Type 10 in the *TechID* field and then press the Tab key.

h. Type sep072021 in the *WODate* field and then press the Tab key. Notice that the placeholder characters and spaces appear as soon as the first letter is typed. Notice also that the first character is converted to uppercase and that the spaces do not need to be typed; Access moves automatically to the next position after the month and day are typed.

i. Type Replace keyboard in the *Descr* field and then press the Tab key.

j. Type sep102021 in the *ServDate* field and then press the Tab key.

k. Complete the remainder of the record as follows:

Hours	.5
Rate	45.50
Parts	67.25
Comments	Serial Number AWQ-982358

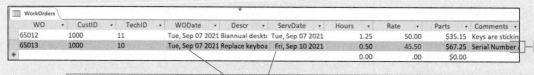

Notice that once the date is accepted into the field, the custom *Format* property controls how the date is presented in the datasheet; the abbreviated day of the week is at the beginning of the field and spaces are between the month, day, and year.

5. Display the datasheet in Print Preview. Change the orientation to landscape. Set the top and bottom margins to 1 inch and the left and right margins to 0.25 inch. Print the datasheet and then close Print Preview.

6. Close the WorkOrders table. Click Yes when prompted to save changes to the table layout.

Check Your Work

Tutorial

Review: Modifying Field Properties in Design View

Other field properties that promote accuracy when entering data and that should be considered when designing database tables include the Default Value, Validation Rule, and Validation Text properties. Use the Default Value property to populate the field in a new record with the field value that is used most often. For example, when most employees live in the same city and state, use default values for these fields to ensure consistent spelling and capitalization within the table. The text appears automatically in the fields when new records are added to the table. The user can accept the default value by pressing the Tab key or the Enter key to move past the field or type new data in the field. In Level 1, default values were created using the Default Value button in the Properties group on the Table Tools Fields tab. In Design view, the Default Value property is located below the Caption property.

Tutorial

Review: Applying a Validation Rule in Design View

Use the Validation Rule and Validation Text properties to enter conditional statements that are checked against new data entered into the field. Invalid entries that do not meet the conditional statement test are rejected. For example, a validation rule on a field used to store labor rates can check that a minimum labor rate value is entered in each record. In Level 1, a validation rule was added in Design view. The Validation Rule and Validation Text properties are located just above the Required property.

Tutorial

Working with a Long Text Field

Quick Steps

Enable Rich Text Formatting in Long Text Field
1. Open table in Design view.
2. Select field defined as Long Text.
3. Double-click in *Text Format* property box.
4. Save table.

Track Changes in Long Text Data Type Field
1. Open table in Design view.
2. Select field defined as Long Text.
3. Double-click in *Append Only* property box.
4. Save table.

Working with a Long Text Data Type Field

By default, Access formats a Long Text data type field as plain text. However, formatting attributes can be applied to text by enabling rich text formatting. Enabling rich text formatting in a Long Text data type field allows the user to change the font, apply bold or italic formatting, or add font color to text, among other formatting options. To enable rich text formatting, change the Text Format property to *Rich Text*.

The Append Only property for a Long Text data type field is set to *No* by default. Change the property to *Yes* to track changes made to the field value in the datasheet. Scroll down the General tab in the *Field Properties* section to locate the Append Only property. When this property is set to *Yes*, Access maintains a history of additions to the field, which can be viewed in the datasheet. Changing the Append Only property to *No* causes Access to delete any existing history.

In Activity 2a, the text format of the *Comments* field is changed from *Plain Text* to *Rich Text* and then bold red formatting is applied to the serial number of the new keyboard. These formatting changes make it easier to find important information in the *Comments* field—in this case, the serial number. Changing the Append Only property from *No* to *Yes* allows the user to keep track of the dates for any comments entered into the field.

Activity 2a Working with Rich Text Formatting and Maintaining a History of Changes in a Long Text Data Type Field

Part 1 of 2

1. With **1-RSRCompServ** open, enable rich text formatting and turn on tracking of history in a field defined as a Long Text data type field by completing the following steps:
   a. Open the WorkOrders table in Design view.
   b. Make *Comments* the active field.
   c. Double-click in the *Text Format* property box (displays *Plain Text*). It should now read *Rich Text*.
   d. At the Microsoft Access message box indicating that the field will be converted to Rich Text, click Yes.

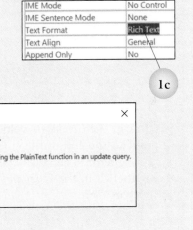

1c

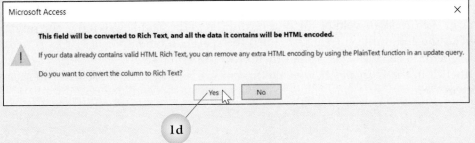

1d

e. If necessary, scroll down the General tab in the *Field Properties* section to locate the *Append Only* property box.

f. Double-click in the *Append Only* property box to change *No* to *Yes*.

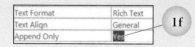

g. Save the table.

2. Switch to Datasheet view.

3. Minimize the Navigation pane and then adjust all the column widths except the *Descr* and *Comments* columns to best fit.

4. Change the column width of the *Comments* field to 25 characters.

5. Select the serial number text *AWQ-982358* in the second record in the *Comments* field and then apply bold formatting and the standard red font color using the buttons in the Text Formatting group on the Home tab. Click at the end of the serial number to deselect the text.

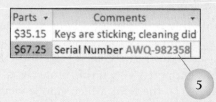

6. Click in the *Comments* field in the first record. Press the End key to move the insertion point to the end of the existing text. Press the spacebar, update the record by typing Microsoft wireless keyboard was recommended., and then press the Enter key to save the changes and move to the next row.

7. Right-click in the *Comments* field of the first record and then click *Show column history* at the shortcut menu.

8. Click OK after reading the text in the History for Comments dialog box.

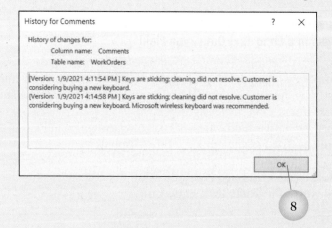

9. Click in the *Comments* field of the first record. Press the End key to move the insertion point to the end of the current text. Press the spacebar, type See work order 65013 for replacement keyboard request., and then press the Enter key.

10. Right-click in the *Comments* field of the first record and then click *Show column history* at the shortcut menu.

11. Click OK after reading the text in the History for Comments dialog box.

12. Display the datasheet in Print Preview. Change to landscape orientation. Set the top and bottom margins to 1 inch and the left and right margins to 0.25 inch. Print the datasheet and then close Print Preview.

13. Close the WorkOrders table. Click Yes if prompted to save changes to the table layout and then redisplay the Navigation pane.

History for Comments

History of changes for:

Column name: Comments
Table name: WorkOrders

[Version: 1/9/2021 4:11:54 PM] Keys are sticking; cleaning did not resolve. Customer is considering buying a new keyboard.
[Version: 1/9/2021 4:14:58 PM] Keys are sticking; cleaning did not resolve. Customer is considering buying a new keyboard. Microsoft wireless keyboard was recommended.
[Version: 1/9/2021 4:21:25 PM] Keys are sticking; cleaning did not resolve. Customer is considering buying a new keyboard. Microsoft wireless keyboard was recommended. See work order 65013 for replacement keyboard request.

OK

11

> Check Your Work

> Tutorial

Creating an Attachment Field

Creating an Attachment Data Type Field and Attaching Files to Records

Use an Attachment data type field to store several files in a single field attached to a record. The attachments can be opened within Access and viewed in the program from which the document originated. For example, attach a Word document to a field in a record. Opening the attached file in the Access table causes Microsoft Word to start and the document to display. A file that is attached to a record cannot be larger than 256 megabytes (MB).

An Attachment data type field displays with a paper clip in Datasheet view. To manage the attached files, double-click the paper clip to open the Attachments dialog box, shown in Figure 1.4. A field that is created with an Attachment data type cannot be changed. Multiple files can be attached to a record, as long as the combined size of all the files does not exceed 2 gigabytes (GB).

Any file created within the Microsoft Office suite can be attached to a record or an image file (.bmp, .jpg, .gif, .png), a log file (.log), a text file (.txt), or a compressed file (.zip). Some files, such as any file ending with *.com* or *.exe*, are considered potential security risks and therefore blocked by Access.

Quick Steps

Create Attachment Field
1. Open table in Design view.
2. Click in first blank field row.
3. Type field name.
4. Click in *Data Type* column.
5. Click option box arrow.
6. Click *Attachment*.
7. Save table.

Attach Files to Record
1. Open table in Datasheet view.
2. Double-click paper clip in record.
3. Click Add button.
4. Navigate to drive and/or folder location.
5. Double-click file name.
6. Click OK.

View Attached File
1. Open table in Datasheet view.
2. Double-click paper clip in record.
3. Double-click file name.
4. View file contents.
5. Exit source program.
6. Click OK.

Figure 1.4 Attachments Dialog Box

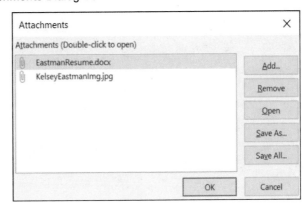

Saving an Attached File to Another Location

To save a copy of an attachment outside Access, select the file and then click the Save As button in the Attachments dialog box. At the Save Attachment dialog box, navigate to the drive and/or folder in which the duplicate copy is to be saved, click the Save button, and then click OK to close the Attachments dialog box.

Editing or Removing an Attached File

Access 365 provides two options for editing an attachment. The first option is to save the attached file to another location, as noted above. After the necessary changes are made, save the file, remove the original attachment from the database, and then attach the edited file. The second option is to open the attachment directly in Access and enable the content. After making the necessary changes in the source program (for example, Word), close the application and save the changes, click OK in the Attachments dialog box and respond to the message asking if you would like to save the updates to the database.

To remove a file attached to a record in a database, open the Attachments dialog box in the record containing the file attachment, click the file name for the file to be removed, click the Remove button, and then click OK to close the Attachments dialog box.

Activity 2b Creating an Attachment Data Type Field, Attaching Files to a Record, and Viewing the Contents of Attached Files

Part 2 of 2

1. With **1-RSRCompServ** open, create a new field in the Technicians table to store file attachments by completing the following steps:
   a. Open the Technicians table in Design view.
   b. Click in the blank row below *CPhone*, type Attachments, and then press the Tab key.
   c. Click the option box arrow in the *Data Type* column and then click *Attachment* at the drop-down list.
   d. Save the table.
2. Switch to Datasheet view.
3. Add the following data in the first row of the datasheet:

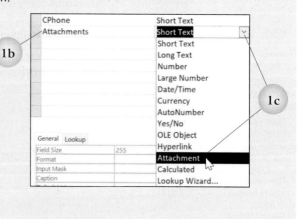

TechID	10
SSN	000-43-5789
FName	Kelsey
LName	Eastman
Street	550 Montclair Street
City	Detroit
State	MI
ZIP	48214-3274
HPhone	"" (Recall that double quotation marks indicate a zero-length field.)
CPhone	313-555-6315

4. Attach two files to the record for Kelsey Eastman by completing the following steps:
   a. Double-click the paper clip in the first row of the datasheet. An Attachment data type field displays a paper clip in each record in a column and has a paper clip in the field name row. The number in brackets next to the paper clip indicates the number of files attached to the record.

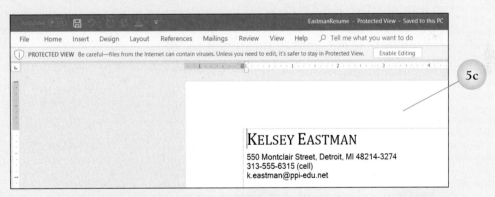

   b. At the Attachments dialog box, click the Add button.
   c. At the Choose File dialog box, navigate to the AL2C1 folder on your storage medium.
   d. Click the file named *EastmanResume.*
   e. Press and hold down the Ctrl key, click the file named *KelseyEastmanImg*, and then release the Ctrl key.
   f. Click the Open button.
   g. Click OK. Access closes the Attachments dialog box and displays *(2)* next to the paper clip in the first record.
5. Open the attached files by completing the following steps:
   a. Double-click the paper clip in the first row of the datasheet to open the Attachments dialog box.
   b. Double-click *EastmanResume.docx* in the *Attachments* list box to open the Word document.
   c. Read the resume in Microsoft Word and then exit Word.

   d. Double-click *KelseyEastmanImg.jpg* in the *Attachments* list box to open the picture file.
   e. View the picture and then exit the photo viewer program.
   f. Click OK to close the Attachments dialog box.
6. Adjust all the column widths to best fit.
7. Display the datasheet in Print Preview. Change the orientation to landscape, print the datasheet, and then close Print Preview.
8. Close the Technicians table. Click Yes when prompted to save changes to the table layout.
9. Close **1-RSRCompServ**.

Check Your Work

Chapter Summary

- Database designers plan the tables for a new database by analyzing sample data, input documents, and output requirements to generate the entire set of data elements needed.

- Once all the data has been identified, the designer maps out the number of tables required.

- Each table holds data only for a single entity (topic). Data is split into the smallest unit that will be manipulated.

- Designers also consider the relationships that will be created. Doing so in the planning stage helps determine if a field needs to be added to a table to join tables.

- Data redundancy should be avoided. This means that fields should not be repeated in another table, except those fields needed to join tables in a relationship.

- A diagram of a database portrays the database tables, providing field names, data types, field sizes, and notation of the primary key fields.

- A field is assigned a data type by selecting a type appropriate for the kind of data that will be accepted into the field.

- Changing the Field Size property is one way to restrict entries in the field to a maximum length. This prevents having longer entries added to the field by accident.

- Add field properties before creating queries, forms, and reports as the properties carry over to these objects.

- Change the Required property to *Yes* to make sure data is typed into the field when a new record is added to the table.

- A zero-length field is entered into a record by typing two double quotation marks with no space between them. This method is used to indicate that a field value does not apply to the current record.

- Disallow zero-length strings by changing the Allow Zero Length property to *No*.

- The Format property controls the display of data in a field. Custom formats can be created by typing the appropriate format codes in the *Format* property box.

- A custom numeric format can contain four sections: one for positive values, one for negative values, one for zero, and one for null values.

- Use an input mask to control the type and pattern of data entered into the field.

- Create a custom input mask for a Short Text or Date/Time data type field by typing the appropriate input mask codes in the *Input Mask* property box.

- A Long Text data type field can be formatted using rich text formatting options in the Text Formatting group on the Home tab. To enable rich text formatting, change the Text Format property to *Rich Text*.

- Change the Append Only property of a Long Text data type field to *Yes* to track changes made to field values.

- An Attachment data type field can be used to store files in a single field attached to a record.

- Double-click the paper clip in the Attachment data type field for a record to add, view, save, or remove a file attachment.

Commands Review

FEATURE	RIBBON TAB, GROUP	BUTTON	KEYBOARD SHORTCUT
create table in Design view	Create, Tables		
minimize Navigation pane		«	F11
redisplay Navigation pane		»	F11
switch to Datasheet view from Design view	Table Tools Design, Views		
switch to Design view from Datasheet view	Home, Views		

Microsoft®

Access®

Building Relationships and Lookup Fields

Performance Objectives

Upon successful completion of Chapter 2, you will be able to:

1 Create and edit relationships between tables, including one-to-many, one-to-one, and many-to-many relationships

2 Define a table with a multiple-field primary key field

3 Create and modify a lookup field to populate records with data from another table

4 Create a lookup field that allows having multiple values in records

5 Create single-field and multiple-field indexes

6 Define the term *normalization*

7 Determine if a table is in first, second, or third normal form

Once the table design has been completed, the next step is to establish relationships and relationship options between tables. This involves analyzing the type of relationship that exists between two tables. Some database designers draw a relationship diagram to depict the primary table and the matching record frequency of the related table. In this chapter, you will create and edit relationships and lookup fields, multiple-field primary key fields, multiple-value fields, and indexes. The concept of database normalization and three forms of normalization will be introduced to complete the examination of database design fundamentals.

 Data Files

Before beginning chapter work, copy the AL2C2 folder to your storage medium and then make AL2C2 the active folder.

The online course includes additional training and assessment resources.

Activity 1 Create and Edit Relationships

4 Parts

You will create relationships and edit relationship options for the tables designed to track work orders for RSR Computer Services and create and print a relationship report.

Building Relationships

Hint Are you unsure whether two tables should be related? Consider whether data will need to be extracted from both tables in the same query, form, or report. If yes, then the tables should be joined in a relationship.

After determining which tables to relate to one another, the next step in designing a database is to examine the types of relationships that exist between the tables. A relationship is based on an association between two tables. For example, in the database created in Chapter 1 for RSR Computer Services, there is an association between the Customers table and the WorkOrders table. A customer is associated with all of his or her work orders involving computer maintenance requests, and each work order is associated with the individual customer who requested the service.

When building relationships, consider the associations between tables and how these associations affect the data that will be entered into the tables. In the database diagram presented in Chapter 1, relationships were shown with lines connecting the common field name between tables. In this chapter, consider the type of relationship that should exist between the tables and the relationship options to use to place restrictions on data entry. Access provides for three types of relationships: one-to-many, one-to-one, and many-to-many. Access Level 1, Chapter 2 addressed one-to-many and one-to-one relationships. This chapter begins by reviewing these two relationship types and then discusses how to establish a many-to-many relationship.

Tutorial

Review: Creating a One-to-Many Relationship

Establishing a One-to-Many Relationship

In the computer service database in Chapter 1, the relationship between the Customers table and WorkOrders table exists because a work order involves computer maintenance for a specific customer. The customer is identified by the customer number stored in the Customers table. In the Customers table, only one record exists per customer. In the WorkOrders table, the same customer number can be associated with several work orders. This means that the relationship between the Customers table and WorkOrders table is a one-to-many relationship. This is the most common type of relationship created in Access.

Relationships

A common field is needed to join the Customers table and WorkOrders table, so the *CustID* field is included in both tables. In the Customers table, *CustID* is the primary key field because it contains a unique identification number for each customer. In the WorkOrders table, *CustID* cannot be the primary key field because the same customer can be associated with several work orders. In the WorkOrders table, *CustID* is a type of field referred to as a *foreign key* field. A foreign key field is included in a table for the purpose of creating a relationship to a field that is a primary key field in another table. The Customers-to-WorkOrders one-to-many relationship is illustrated in Figure 2.1.

Figure 2.1 One-to-Many Relationship between the Customers Table and WorkOrders Table

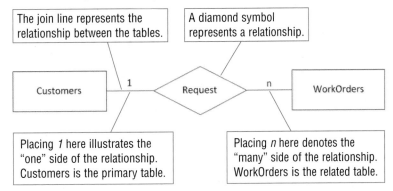

The join line represents the relationship between the tables.

A diamond symbol represents a relationship.

Customers — 1 — Request — n — WorkOrders

Placing *1* here illustrates the "one" side of the relationship. Customers is the primary table.

Placing *n* here denotes the "many" side of the relationship. WorkOrders is the related table.

Quick Steps

Create One-to-Many Relationship
1. Click Database Tools tab.
2. Click Relationships button.
3. Add tables from Show Table dialog box.
4. Close Show Table dialog box.
5. Drag primary key field name from primary table to foreign key field name in related table.
6. Click Create button.

When diagramming a database, designers may choose to show relationships in a separate illustration. In the diagram shown in Figure 2.1, table names are displayed in rectangles and connected with lines to a diamond symbol, which represents a relationship. Inside the diamond, a word (usually a verb) describes the action that relates the two tables. For example, in the relationship shown in Figure 2.1, the word *Request* is used to show that "Customers *request* WorkOrders." On the line connecting the rectangle to the diamond symbol (called the *join line*), a *1* is placed next to the primary table, or the "one" side of the relationship, and an *n* is placed next to the related table, or the "many" side of the relationship.

Activity 1a Creating a One-to-Many Relationship

Part 1 of 4

1. Open **2-RSRCompServ**. This database has the same structure as the database created in Chapter 1. However, additional field properties have been defined and several records have been added to each table to provide data for testing relationships and lookup fields.
2. If a security warning appears in the message bar to indicate that some active content has been disabled, click the Enable Content button.
3. Create a one-to-many relationship between the Customers table and WorkOrders table by completing the following steps:
   a. Click the Database Tools tab.
   b. Click the Relationships button in the Relationships group.
   c. At the Show Table dialog box with the Tables tab active and *Customers* selected in the list box, press and hold down the Ctrl key, click *WorkOrders*, click the Add button, and then release the Ctrl key.
   d. Click the Close button to close the Show Table dialog box.
   e. Resize both table field list boxes by dragging down the bottom borders of the boxes until all the field names are visible.

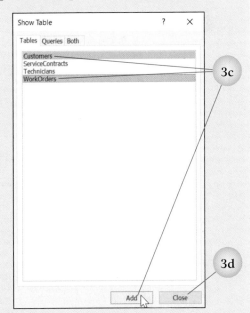

f. Drag the *CustID* field from the Customers table field list box to the *CustID* field in the WorkOrders table field list box. Be careful to drag the common field name from the primary table (Customers) to the related table (WorkOrders) and not vice versa.

g. At the Edit Relationships dialog box, notice that *One-To-Many* appears in the *Relationship Type* section. Access detected this type of relationship because the field used to join the tables is a primary key field in only one of the tables. Establishing this relationship makes *CustID* a foreign key field in the WorkOrders table. Always check that the correct table and field names are shown in the *Table/Query* and *Related Table/Query* option boxes. If the table name and/or the common field name is not shown correctly, click the Cancel button. Errors can occur if the mouse is dragged starting or ending at the wrong table or field. If this happens, click the Cancel button, return to Step 3f and then try again.

h. Click the Create button.

4. Click the Close button in the Relationships group on the Relationship Tools Design tab.

5. Click Yes at the message box asking if you want to save changes to the layout of the Relationships window.

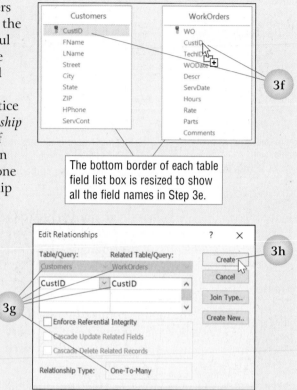

The bottom border of each table field list box is resized to show all the field names in Step 3e.

Check Your Work

Another one-to-many relationship exists between the Technicians table and WorkOrders table. A technician is associated with each work order assigned to that technician and a work order is associated with the technician that carried out the service request. The Technicians-to-WorkOrders relationship diagram is shown in Figure 2.2.

Figure 2.2 One-to-Many Relationship between the Technicians Table and WorkOrders Table

1. With **2-RSRCompServ** open, display the Relationships window by clicking the Database Tools tab and then clicking the Relationships button.
2. Click the Show Table button in the Relationships group on the Relationship Tools Design tab.
3. Double-click *Technicians* in the list box at the Show Table dialog box with the Tables tab selected. Move the Show Table dialog box, if necessary, to verify that the Technicians table has been added and then click the Close button.
4. Drag the bottom and right borders of the Technicians table field list box until all the field names are fully visible.
5. Drag the *TechID* field from the Technicians table field list box to the *TechID* field in the WorkOrders table field list box.
6. Check that the correct table and field names appear in the *Table/Query* and *Related Table/Query* option boxes. If necessary, click the Cancel button and repeat Step 5.
7. Click the Create button.
8. Click the Close button on the Relationship Tools Design tab.
9. Click Yes at the message box asking if you want to save changes to the layout of the Relationships window.

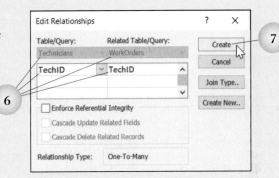

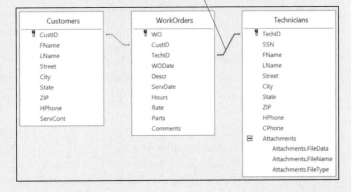

This join line represents the one-to-many relationship created between the Technicians table and the WorkOrders table in Steps 1–7.

Editing a Relationship

Quick Steps

Edit Relationship
1. Click Database Tools tab.
2. Click Relationships button.
3. Click join line between tables.
4. Click Edit Relationships button.
5. Select options.
6. Click OK.

At the Edit Relationships dialog box, shown in Figure 2.3, select relationship options and/or specify the type of join to create. To open the Edit Relationships dialog box, click the Database Tools tab, click the Relationships button, click the join line for the relationship to be edited, and then click the Edit Relationships button.

Once the Edit Relationships dialog box is open, select any of the options. The *Cascade Update Related Fields* and *Cascade Delete Related Records* options do not become active unless referential integrity is turned on. To turn on referential integrity, click the *Enforce Referential Integrity* check box to insert a check mark.

Edit Relationships

Figure 2.3 Edit Relationships Dialog Box

Selecting *Enforce Referential Integrity* places restrictions on data entry. In this example, it means that a technician cannot be assigned to a work order if the technician does not exist in the Technicians table.

Edit Relationships ? ✕

Table/Query:	Related Table/Query:
Technicians ⌄	WorkOrders ⌄
TechID ⌄	TechID ⌃
	⌄

☐ Enforce Referential Integrity
☐ Cascade Update Related Fields
☐ Cascade Delete Related Records

Relationship Type: One-To-Many

OK
Cancel
Join Type..
Create New..

💡 **Hint** To enable referential integrity, the primary key and foreign key fields must be the same data type.

If an error message displays when you are attempting to enforce referential integrity, open each table in Design view and compare the data types for the fields used to join the tables.

Activating referential integrity in a one-to-many relationship is a good way of ensuring that orphan records do not occur. An orphan record is a record in a related table for which no "parent" record exists in the primary table. Assigning a technician to a work order in the WorkOrders table when there is no matching technician record in the Technicians table results in creating an orphan record in the WorkOrders table. Once referential integrity has been turned on, Access checks for the existence of a matching record in the primary table as each new record is added to the related table. If no match is found, Access does not allow the record to be saved.

For example, suppose that referential integrity has not been activated and a typing mistake is made that causes the accidental entry of an unassigned *TechID* to a work order. If the customer later has a question for the technician about the service, no one will know which technician to contact. However, if referential integrity is activated before the record is saved, Access will check the Technicians table and verify that the *TechID* exists. If the *TechID* does not exist, the error message shown in Figure 2.4 will display. Click OK at the message and then enter a *TechID* that exists in the Technicians table.

When the *Enforce Referential Integrity* check box is clicked to insert a check mark, the *Cascade Update Related Fields* and *Cascade Delete Related Records* check boxes become available. When a check mark is inserted in the *Cascade Update Related Fields* check box, Access automatically updates all the occurrences of the same data in the foreign key field in the related table when a change is made to the primary key field in the primary table. When a check mark is inserted in the *Cascade Delete Related Records* check box, deleting a record from the primary table for which related records exist in the related table results in all the related records being automatically deleted.

Join types and situations in which changing the join type is warranted will be discussed in Chapter 3.

Figure 2.4 Referential Integrity Error Message

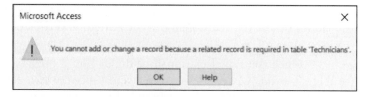

Microsoft Access ✕

⚠ You cannot add or change a record because a related record is required in table 'Technicians'.

OK   Help

1. With **2-RSRCompServ** open, edit the one-to-many relationship between the Customers table and WorkOrders table to enforce referential integrity and add both cascade options by completing the following steps:
   a. Open the Relationships window.
   b. Click to select the join line between the Customers table and WorkOrders table.
   c. Click the Edit Relationships button in the Tools group on the Relationship Tools Design tab.

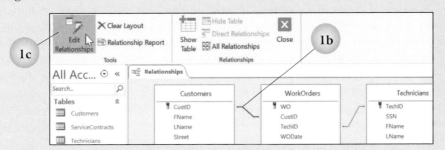

   d. At the Edit Relationships dialog box, click to insert check marks in the *Enforce Referential Integrity* check box, the *Cascade Update Related Fields* check box, and the *Cascade Delete Related Records* check box.
   e. Click OK. The *1* at the primary table (the "one") side of the join line and the infinity symbol (∞) at the related table (the "many" side) of the join line indicate that referential integrity has been activated.

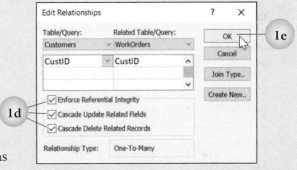

2. Edit the one-to-many relationship between the Technicians table and WorkOrders table to enforce referential integrity and add both cascade options by completing the following steps:
   a. Double-click the join line between the Technicians table and WorkOrders table in the Relationships window. (Or right-click the join line and then click *Edit Relationship* at the shortcut menu.)
   b. At the Edit Relationships dialog box, click to insert check marks in the *Enforce Referential Integrity* check box, the *Cascade Update Related Fields* check box, and the *Cascade Delete Related Records* check box.
   c. Click OK.
3. Notice with referential integrity turned on, the 1 and the infinity symbol (∞) display on the join lines.
4. Close the Relationships window.

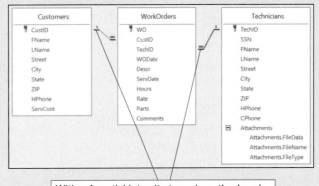

With referential integrity turned on, the 1 and the infinity symbol (∞) display on the join lines.

Check Your Work

Tutorial

Review: Creating
a One-to-One
Relationship

Quick Steps

Create One-to-One Relationship
1. Click Database Tools tab.
2. Click Relationships button.
3. Add tables from Show Table dialog box.
4. Close Show Table dialog box.
5. Drag primary key field name from primary table to primary key field name in related table.
6. Select relationship options.
7. Click Create button.

 Relationship Report

Creating a One-to-One Relationship

In the database for RSR Computer Services, a table is used to store service contract information for each customer. That table, ServiceContracts, is associated with the Customers table. Only one record exists for a customer in the Customers table and each customer subscribes to only one service contract in the ServiceContracts table. This means that the two tables have a one-to-one relationship, as shown in Figure 2.5.

When a new customer is added to the database, the customer's name and contact information are entered into the Customers table first and then the service contract information (including start date, end date, and fee paid) is entered into the ServiceContracts table. When creating the relationship, drag the primary key field name from the table in which the data is entered into the database first (Customers table) to the primary key field name from the table in which the data is entered second (ServiceContracts table). When the relationship is established, the Customers table is placed in the *Table/Query* option box and the ServiceContracts table is placed in the *Related Table/Query* option box; otherwise, when data is being entered, an error message will appear, as shown in Figure 2.4.

To print the relationship, first create the Relationship Report by clicking the Relationship Report button in the Tools group. When Access displays the report in Print Preview, click the Print button on the Print Preview tab.

Figure 2.5 One-to-One Relationship between the Customers Table and ServiceContracts Table

Activity 1d Creating a One-to-One Relationship   Part 4 of 4

1. With **2-RSRCompServ** open, create a one-to-one relationship between the Customers table and ServiceContracts table by completing the following steps:
   a. Open the Relationships window.
   b. Click the Show Table button.
   c. Double-click *ServiceContracts* in the list box at the Show Table dialog box with the Tables tab selected and then click the Close button.
   d. Drag the *CustID* field from the Customers table field list box to the *CustID* field in the ServiceContracts table field list box.
   e. At the Edit Relationships dialog box, check that the correct table and field names appear in the *Table/Query* and *Related Table/Query* option boxes. If necessary, click the Cancel button and repeat Step 1d.
   f. Notice that *One-To-One* appears in the *Relationship Type* section. Access detected this type of relationship because the field used to join the tables is a primary key field in both tables.
   g. Click to insert check marks in the *Enforce Referential Integrity* check box, the *Cascade Update Related Fields* check box, and the *Cascade Delete Related Records* check box.
   h. Click the Create button.

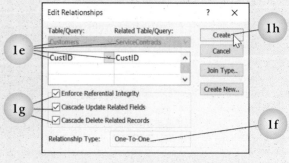

2. Drag the Title bar of the ServiceContracts table field list box to the approximate location in the Relationships window shown below. This makes it easier to view the join line and the *1* at each end of the join line between the Customers and ServiceContracts table field list boxes.

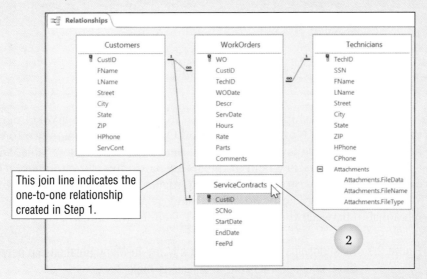

This join line indicates the one-to-one relationship created in Step 1.

3. Create a relationship report by clicking the Relationship Report button in the Tools group on the Relationship Tools Design tab.
4. Access displays the report in Print Preview. Click the Print button in the Print group on the Print Preview tab and then click OK at the Print dialog box.
5. Close the relationship report for **2-RSRCompServ** by clicking the Close Relationship button on the far right side of the screen. At the Microsoft Access message box, click Yes to save the report and click OK at the Save As dialog box to accept the default name.
6. Close the Relationships window.

Check Your Work

Tutorial

Creating a Many-to-Many Relationship

Creating a Many-to-Many Relationship

Consider the association between the Customers table and Technicians table in the RSR Computer Services database. Over time, any individual customer can have computer service work done by many different technicians and any individual technician can perform computer service work at many different customer locations. In other words, a record in the Customers table can be matched to many records in the Technicians table and a record in the Technicians table can be matched to many records in the Customers table. This is an example of a many-to-many relationship. A diagram of the many-to-many relationship between the Customers table and Technicians table is shown in Figure 2.6.

A many-to-many relationship is problematic because it creates duplicate records. If the same customer number is associated with many technicians and vice versa, many duplicates will occur in the two tables and Access may experience data conflicts when trying to identify unique records. To resolve the duplication and create unique entries, a third table is used to associate, or link, the many-to-many tables. That table is called a *junction table* and contains the primary key fields from both tables in the many-to-many relationship as its foreign key fields. Using the junction table, two one-to-many relationships are created.

Figure 2.6 Many-to-Many Relationship between the Customers Table and Technicians Table

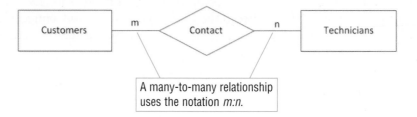

A many-to-many relationship uses the notation *m:n*.

In the Relationships window shown in Figure 2.7, the WorkOrders table is the junction table. Notice that the WorkOrders table contains two foreign key fields: *CustID*, which is the primary key field in the Customers table, and *TechID*, which is the primary key field in the Technicians table. One-to-many relationships exist between the Customers and WorkOrders tables and the Technicians and WorkOrders tables. These two one-to-many relationships create a many-to-many relationship between the Customers and Technicians tables.

Figure 2.7 Relationships Window Showing a Many-to-Many Relationship between the Customers Table and Technicians Table

The Customers and Technicians tables have a many-to-many relationship via the WorkOrders junction table.

The WorkOrders junction table contains two foreign key fields: the primary key field from each table in the many-to-many relationship.

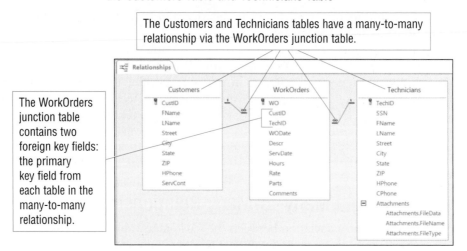

You will create a new table that requires three fields to uniquely identify each record. To restrict data entry in the table, you will create fields that each display a list from which the user selects the field value(s).

Defining a Multiple-Field Primary Key Field

Quick Steps

Define Multiple-Field Primary Key Field
1. Open table in Design view.
2. Select first field.
3. Press and hold down Shift (adjacent row) or Ctrl (nonadjacent row) and select second field.
4. Click Primary Key button.
5. Save table.

Primary Key

💡 **Hint** Delete a primary key field designation by opening the table in Design view, activating the primary key field, and then clicking the Primary Key button.

Defining a Multiple-Field Primary Key Field

In most tables, one field is designated the primary key field. However, in some situations, a single field may not be guaranteed to hold unique data. Look at the fields in the table shown in Figure 2.8. This is a new table that will be created in the RSR Computer Services database to store computer profiles for RSR customers. The company stores the profiles as a service to clients, who sometimes forget their login information. Technicians can also access the profile data when troubleshooting at a customer's site.

A customer may have more than one computer in his or her home or office and each computer may have a different profile for each username. The *CustID* field will not serve as the primary key field if the customer has more than one record in the Profiles table. However, a combination of the three fields *CustID, CompID,* and *Username* will uniquely identify each record. In this table, three fields will be defined as primary key fields. A primary key field that is made up of two or more fields is called a *composite key* field.

Figure 2.8 Activity 2a Profiles Table

Profiles		
*CustID	Short Text	4
*CompID	Short Text	2
*Username	Short Text	15
Password	Short Text	15
Remote	Yes/No	

Activity 2a Creating a New Table with a Multiple-Field Primary Key   Part 1 of 5

1. With **2-RSRCompServ** open, create a new table to store customer profiles by completing the following steps:
   a. Click the Create tab and then click the Table Design button in the Tables group.
   b. Type the field names, assign the data types, and change the field sizes according to the data structure shown in Figure 2.8.

2. Hover the mouse pointer over the field selector bar (the blank column left of the field names) next to *CustID* until the pointer changes to a black right-pointing arrow and then click to select the field.

3. Press and hold down the Shift key, click in the field selector bar next to *Username*, and then release the Shift key. The three adjacent fields *CustID*, *CompID*, and *Username* are now selected.

4. Click the Primary Key button in the Tools group on the Table Tools Design tab. Access displays the primary key icon next to each field name.

5. Click in any data type field to deselect the first three rows.

6. Save the table with the name *Profiles*.

7. Close the table.

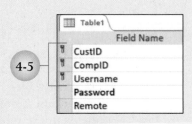

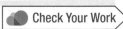

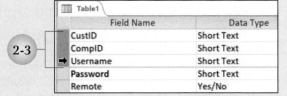
Check Your Work

 Tutorial

Creating a Field to
Look Up Values in
Another Table

 Tutorial

Modifying Lookup
List Properties

Quick Steps

Create Lookup Field to Another Table

1. Open table in Design view.
2. Click in column of lookup field.
3. Click option box arrow.
4. Click *Lookup Wizard*.
5. Click Next.
6. Choose table and click Next.
7. Choose fields to display in column.
8. Click Next.
9. Choose field by which to sort.
10. Click Next.
11. If necessary, expand column widths.
12. Clear *Hide key column*.
13. Click Next.
14. Choose field value to store in table.
15. Click Next.
16. Click Finish.
17. Click Yes.

Creating a Field to Look Up Values in Another Table

In Level 1, Chapter 4, the Lookup Wizard was used to create a lookup field where the values in the list were typed. The Lookup Wizard can also be used to create a lookup field to look up values found in records from another table. Lookup fields allow the user to enter data by pointing and clicking, rather than by typing the entry in the field.

A lookup field that draws its data from a field in another table can be useful in several ways. Data can be restricted to items within the list, which prevents orphan records, data entry errors, and spelling inconsistencies from occurring. The lookup field can also provide more information to help the user select the correct option. For example, suppose that a lookup field requires the user to select a customer's identification number. If the lookup field displays a drop-down list of identification numbers, identifying the correct number will be difficult. However, if the lookup field displays the identification number along with the customer's name, the correct entry will be easy to identify. When the user chooses the field entry based on the name, Access automatically enters the correct identification number.

When using the Lookup Wizard to create a lookup field, make sure to do so before the relationships are created. If a relationship already exists between the table for the lookup field and the source data table, Access will display a message stating that the relationship needs to be deleted before the Lookup Wizard can run.

Use the Lookup tab in the *Field Properties* section to change the *Limit to List* property from *No* to *Yes*. A user will not be able to type in an entry that is not in the list.

1. With **2-RSRCompServ** open, open the Profiles table in Design view.
2. Create a lookup field to select and enter a customer's identification number from a list of customers in the Customers table by completing the following steps:
   a. With *CustID* the active field, click in the *Data Type* column, click the option box arrow, and then click *Lookup Wizard* at the drop-down list.
   b. At the first Lookup Wizard dialog box with *I want the lookup field to get the values from another table or query* selected, click the Next button.
   c. At the second Lookup Wizard dialog box with *Table: Customers* already selected in the *Which table or query should provide the values for your lookup field?* list box, click the Next button.

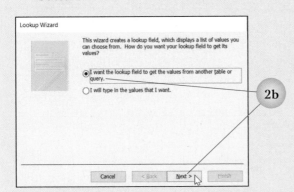
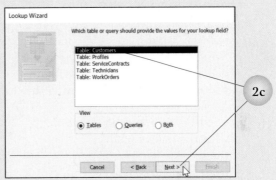

   d. At the third Lookup Wizard dialog box, double-click *FName* in the *Available Fields* list box to move the field to the *Selected Fields* list box.
   e. Double-click *LName* in the *Available Fields* list box to move the field to the *Selected Fields* list box and then click the Next button.

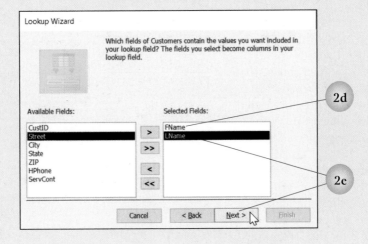

f. At the fourth Lookup Wizard dialog box, click the first sort option box arrow, and then click *LName* at the drop-down list. Notice that up to four sort keys can be defined to sort the lookup list and that an Ascending button appears next to each *Sort* option box. You can change the sort order from Ascending to Descending by clicking the Ascending button. Click the Next button.

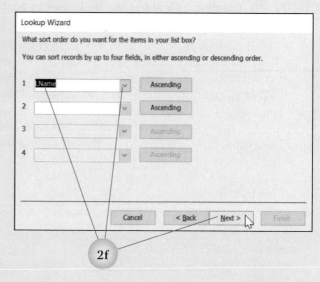

g. At the fifth Lookup Wizard dialog box, expand the column widths if necessary to display all the data. Scroll down the list of entries in the dialog box. Notice that the column widths are sufficient to show all the text.

h. To view the customer identification numbers with the names while the list is open in a record, click the *Hide key column (recommended)* check box to remove the check mark. Removing the check mark displays the *CustID* field values as the first column in the lookup list.

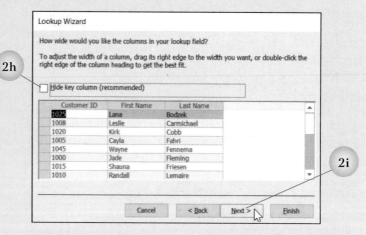

i. Click the Next button.

j. At the sixth Lookup Wizard dialog box, choose the field value to be stored in the table when an entry is selected in the drop-down list. With *CustID* already selected in the *Available Fields* list box, click the Next button.

k. At the last Lookup Wizard dialog box, click the Finish button to accept the existing field name for the lookup field of *CustID*.

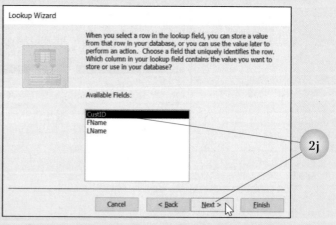

l. At the Lookup Wizard message box stating that the table must be saved before relationships can be created, click Yes to save the table. Access automatically creates a relationship between the Customers table and Profiles table based on the *CustID* field used to create the lookup field.

3. Close the Profiles table.

1. With **2-RSRCompServ** open, open the Profiles table in Design view.
2. Type the following text in the *Caption* property box for each of the following fields:
   - *CustID* — Customer ID
   - *CompID* — Computer ID
   - *Remote* — Remote Access?
3. Modify the lookup field properties to restrict entries in new records to items within the list by completing the following steps:
   a. Make *CustID* the active field.
   b. Click the Lookup tab in the *Field Properties* section.
   c. Look at the entries in all the property boxes for the Lookup tab. These entries were created by the Lookup Wizard.
   d. Double-click in the *Limit To List* property box to change *No* to *Yes*. This means that the field will accept data only from existing customer records. A user will not be able to type in an entry that is not in the list.

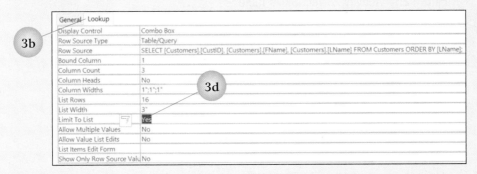

   e. Save the table.
4. Switch to Datasheet view.
5. With *Customer ID* in the first row of the datasheet as the active field, click the option box arrow in the field and then click *Jade Fleming* at the drop-down list. Notice that Access inserts *1000* as the field value in the first column since that is the customer number associated with the selected name.

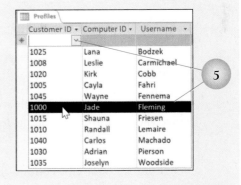

6. Type the remaining data as indicated:
   - *Computer ID* — D1
   - *Username* — jade
   - *Password* — P$ck7
   - *Remote Access?* (Leave blank to indicate *No*.)
7. Best fit the width of each column.
8. Print and then close the Profiles datasheet. Click Yes when prompted to save changes to the table layout.

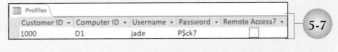

Check Your Work

Creating a Field That Allows Multiple Values

The industry certifications that a technician has achieved could be organized by creating a separate field for each certification. A more efficient approach is to create a single field that displays a list of certifications with check boxes to indicate whether they have been attained.

Look at the fields in the table structure shown in Figure 2.9. For each technician, a list in the *Certifications* field is opened and the check box next to the applicable certification title is checked if the technician has obtained that specific certification. In the *OperatingSys* field, another list can be used to keep track of the operating systems for which the technician is considered an expert.

Use the Lookup Wizard to create a field to store multiple values. Choose to look up the values in a field in another table or create a custom value list. At the last Lookup Wizard dialog box, make sure to click the *Allow Multiple Values* check box to insert a check mark.

♀Hint Multiple-value fields should be used when more than one choice from a small list needs to be stored. This makes it unnecessary to create an advanced database design.

Ǫuick Steps

Create Multiple-Value Lookup Field
1. Open table in Design view.
2. Start Lookup Wizard for desired field.
3. Create list by typing values or binding data to field in another table.
4. At last Lookup Wizard dialog box, insert check mark in *Allow Multiple Values* check box.
5. Click Finish.
6. Click Yes.

Figure 2.9 Activity 2d TechSkills Table

TechSkills		
*TechID	Short Text	2
Certifications	Short Text	20
OperatingSys	Short Text	20
NetworkSpc	Yes/No	
WebDesign	Yes/No	
Programming	Yes/No	

Activity 2d Creating Fields That Allow Multiple Values in a New Table

Part 4 of 5

1. With **2-RSRCompServ** open, create a new table to store technician competencies by completing the following steps:
   a. Create a new table using Design view.
   b. Type the field names and assign the data types according to the data structure shown in Figure 2.9.
   c. Assign the field denoted with an asterisk in Figure 2.9 as the primary key field.
   d. Save the table with the name *TechSkills*.
2. Create a lookup field to select a technician from a list of names in the Technicians table by completing the following steps:
   a. Click in the *Data Type* column for the *TechID* field, click the option box arrow that appears, and then click *Lookup Wizard*.
   b. Click the Next button at the first Lookup Wizard dialog box.
   c. Click *Table: Technicians* and then click the Next button.
   d. Double-click *FName* in the *Available Fields* list box to move the field to the *Selected Fields* list box.
   e. Double-click *LName* in the *Available Fields* list box and then click the Next button.
   f. Sort by *LName* and then click the Next button.

g. With a check mark in the *Hide key column (recommended)* check box, click the Next button to accept the current column widths. (In this lookup example, you are electing not to show the technician's ID field value. Although you will view and select by name, Access stores the primary key field's value in the table. *TechID* is considered the bound field and *FName* and *LName* are considered display fields).

h. Click the Finish button and then click Yes to save the table.

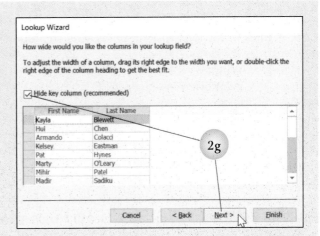

3. Create a lookup field that allows multiple values for certification information by completing the following steps:

a. Click in the *Data Type* column for the *Certifications* field, click the option box arrow that appears, and then click *Lookup Wizard*.

b. Click *I will type in the values that I want* and then click the Next button.

c. At the second Lookup Wizard dialog box, type the following entries in *Col1* pressing the Tab key to move down to the next entry:
   CCNA Cloud
   CCNA Wireless
   Comp TIA A+
   Microsoft MCT
   MOS Master

d. Click the Next button.

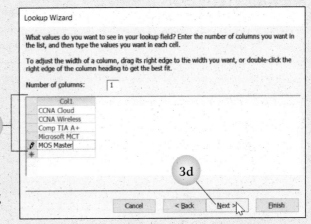

e. At the last Lookup Wizard dialog box, click the *Allow Multiple Values* check box to insert a check mark and then click the Finish button.

f. At the message box indicating that once the field is set to store multiple values, the action cannot be undone, click Yes to change the *Certifications* field to multiple values.

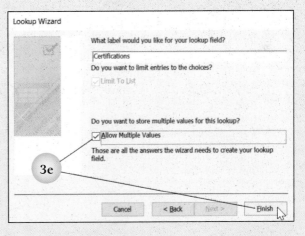

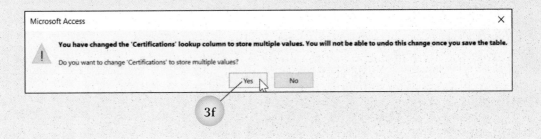

4. Complete steps similar to those in Steps 3a–3f to create a lookup list to store multiple values in the *OperatingSys* field using the following value list:
> Windows 10
> Windows 8
> Windows 7
> Linux
> Unix

5. Save and then close the TechSkills table.

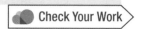

Activity 2e Assigning Multiple Values in a Lookup List

Part 5 of 5

1. With **2-RSRCompServ** open, open the TechSkills table in Design view.
2. Type the following text in the *Caption* property box for each field as indicated:

TechID	Technician ID
OperatingSys	Operating Systems
NetworkSpc	Network Specialist?
WebDesign	Design Web Sites?
Programming	Programming?

3. Save the table and then switch to Datasheet view.
4. Add a new record to the table by completing the following steps:
   a. With the insertion point positioned in the *Technician ID* column, click the option box arrow and then click *Kelsey Eastman* at the drop-down list. Notice that Access displays the technician's first name in the column. *FName* is considered a display field for this column but the identification number associated with the name *Kelsey Eastman* is stored in the table.
   b. Press the Tab key and then click the option box arrow in the *Certifications* column.
   c. Since *Certifications* is a multiple-value field, the drop-down list displays with check boxes next to all the items. Click the *CCNA Cloud* check box and the *Microsoft MCT* check box to insert check marks and then click OK.
   d. Press the Tab key and then click the option box arrow in the *Operating System* column.
   e. Click the *Windows 10, Windows 8,* and *Linux* check boxes to insert check marks and then click OK.
   f. Press the Tab key and then press the spacebar to insert a check mark in the *Network Specialist?* check box.
   g. Press the Tab key three times to finish the record, leaving the check boxes blank in the *Design Web Sites?* and *Programming?* columns.
5. Best fit the width of each column.
6. Print the TechSkills table in landscape orientation with left and right margins of 0.25 inch.
7. Close the TechSkills table. Click Yes when prompted to save changes to the layout.

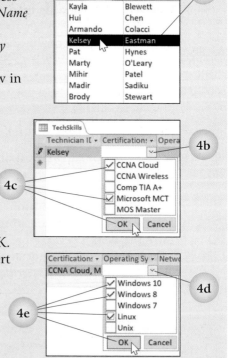

Check Your Work

Tutorial

Creating an Index

Quick Steps

Create Single-Field Index
1. Open table in Design view.
2. Make field active.
3. Click in *Indexed* property box.
4. Click option box arrow.
5. Click *Yes (Duplicates OK)* or *Yes (No Duplicates)*.
6. Save table.

Create Multiple-Field Index
1. Open table in Design view.
2. Click Indexes button.
3. Click in first blank row in *Index Name* column.
4. Type name for index.
5. Press Tab.
6. Click option box arrow in *Field Name* column.
7. Click field.
8. If necessary, change sort order.
9. Click in *Field Name* column in next row.
10. Click option box arrow.
11. Click field.
12. If necessary, change sort order.
13. Repeat Steps 9–12 until finished.
14. Close Indexes window.

  Indexes

💡 **Hint** An index cannot be generated for fields with a data type of OLE Object or Attachment.

Creating an Index

An index is a list created by Access containing pointers that direct Access to the locations of specific records in a table. A database index is very similar to an index found at the end of a book. Search the book's index for a keyword associated with the topic being searched, and the index gives the page number(s) in the book that contains information on that topic. Using a book's index allows information to be retrieved quickly and efficiently. Although the information in an Access index cannot be seen, it operates in much the same way, reducing the amount of time it takes to find a particular record.

Access automatically generates an index for a field designated the primary key field in a table. In a database with a large number of records, it is useful to identify fields other than the primary key field that are often sorted or searched and to create indexes for these fields to speed up sorting and searching. For example, in the Customers table in the RSR Computer Services database, creating an index for the *LName* field is a good idea because the table data will frequently be sorted by a customer's last name.

An index can be created to restrict the data in a field to unique values. This creates a field similar to a primary key field in that Access will not allow two records to hold the same data. For example, an email field in a table that is frequently searched is a good candidate for an index. To avoid data entry errors in a field that should contain unique values (and is not the primary key field), set up the index so it will not accept duplicates.

Create a multiple-field index if a large table is frequently sorted by two or more fields at the same time. In Table Design view, click the Indexes button to open the Indexes window, as shown in Figure 2.10. Create an index for a combination of up to 10 fields.

Figure 2.10 Indexes Window

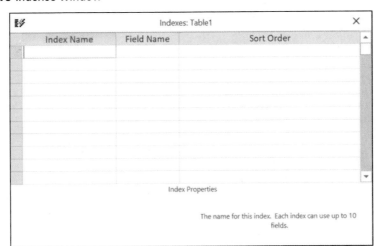

1. With **2-RSRCompServ** open, open the Customers table in Design view.
2. Create a single-field index for the *ZIP* field by completing the following steps:
   a. Make *ZIP* the active field.
   b. Double-click in the *Indexed* property box to change *No* to *Yes (Duplicates OK)*.
   c. Save the table.
3. Create a multiple-field index for the *LName* and *FName* fields by completing the following steps:
   a. Click the Indexes button in the Show/Hide group on the Table Tools Design tab.
   b. At the Indexes: Customers window, click in the first blank row in the *Index Name* column (below *ZIP*) and then type Names.
   c. Press the Tab key, click the option box arrow that appears in the *Field Name* column, and then click *LName* at the drop-down list. The sort order for *LName* defaults to *Ascending*.

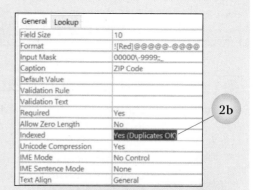

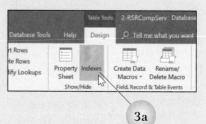

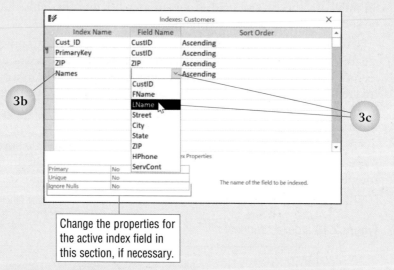

Change the properties for the active index field in this section, if necessary.

   d. In the *Field Name* column, click in the row below *LName*, click the option box arrow that appears, and then click *FName*.
   e. Close the Indexes: Customers window.
   f. Save the table.
4. Close the Customers table.
5. Close **2-RSRCompServ**.

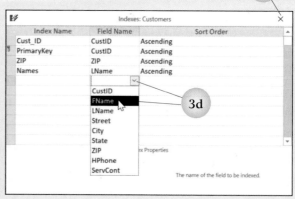

Normalizing a Database

Normalizing a database involves reviewing the database structure to ensure that the tables are set up to eliminate redundancy. If data redundancy is discovered, the process of normalization often involves splitting fields into smaller units and/or breaking down larger tables into smaller tables and creating relationships to remove repeating groups of data. Three normalization states are tested: first normal form, second normal form, and third normal form.

Checking First Normal Form

A table meets first normal form when it does not contain any fields that can be broken into smaller units and when it does not have similar information stored in several fields.

For example, a table that contains a single field called *TechnicianName* that stores the technician's first and last names in the same column violates first normal form. To correct the structure, split *TechnicianName* into two fields, such as *TechLastName* and *TechFirstName*.

A table that has multiple fields and in which each field contains similar data—such as *Week1*, *Week2*, *Week3*, and *Week4*—also violates first normal form. To correct this structure, delete the four week fields and replace them with a single field named *WeekNumber*.

Checking Second Normal Form

Meeting second normal form is of concern only for a table that has a multiple-field primary key field (composite key field). A table with a composite key field meets second normal form when it is in first normal form and when all of its fields are dependent on all the fields that form the primary key field.

Hint If a field cannot relate to the individual fields of a composite key field, then it should be included in a different table.

For example, assume that a table is defined with two fields that form the primary key field: *CustID* and *ComputerID*. Also assume that a field in the same table is titled *EmailAdd*. The contents of the *EmailAdd* field are dependent on the customer only (not the computer). Since *EmailAdd* is not dependent on both *CustID* and *ComputerID*, the table is not in second normal form. To correct the structure, delete the *EmailAdd* field. (The *EmailAdd* field belongs in another table in the database.)

Checking Third Normal Form

Meeting third normal form applies to a table that has a single primary key field and is in first normal form. If a field exists in the table for which the field value is not dependent on the field value of the primary key field, the table is not in third normal form.

For example, assume that a table is defined with a single primary key field titled *TechnicianID*. Also assume that fields in the same table are titled *PayCode* and *PayRate*. Finally, assume that a technician's pay rate is dependent on the pay code assigned to him or her. Since the pay rate is dependent on the field value in the pay code field and not on the technician's identification number, the table is not in third normal form. To convert the table to third normal form, delete the *PayRate* field from the table. (The *PayRate* field belongs in another table in the database.)

Chapter Summary

- When building relationships, consider the frequency of matching data in the common field in both tables to determine if the relationship is one-to-many, one-to-one, or many-to-many.

- The most common type of relationship is one-to-many. It involves joining tables by dragging the primary key field from the "one" table to the foreign key field in the "many" table.

- In a one-to-many relationship, only one record for a matching field value exists in the primary table, while many records for the same value can exist in the related table.

- A relationship diagram depicts the two tables joined in the relationship and the type of relationship between them.

- Enforce referential integrity to place restrictions on new data entered into a related table at the Edit Relationships dialog box. A record is not allowed in the related table if a matching record does not already exist in the primary table.

- Selecting the *Cascade Update Related Fields* option automatically updates all the occurrences of the same data in the foreign key field in the related table when a change is made to the primary key field in the primary table.

- Selecting the *Cascade Delete Related Records* option automatically deletes related records when a record is deleted from the primary table.

- In a one-to-one relationship, only one record exists that has a matching value in the joined field in both tables.

- In a many-to-many relationship, many records can exist that have matching values in the joined fields in both tables.

- A junction table is used to create a many-to-many relationship. A junction table contains a minimum of two fields, which are the primary key fields from the two tables in the many-to-many relationship.

- Using a junction table, two one-to-many relationships are joined to form a many-to-many relationship.

- In some tables, two or more fields are used to create the primary key field if a single field is not guaranteed to hold unique data.

- A primary key field that is made up of two or more fields is called a *composite key* field.

- A lookup field displays a drop-down list in a field, in which the user points and clicks to enter the field value. The list can be generated from records in a related table or by typing in a value list.

- Once a lookup field has been created, options at the Lookup tab in the *Field Properties* section in Table Design view can be selected to modify individual properties.

- Clicking *Yes* at the Limit To List property box allows entries in the field to be restricted to items within the lookup list.

- A field that allows selecting multiple entries from a drop-down list can be created by clicking *Allow Multiple Values* at the final Lookup Wizard dialog box.

- Access displays check boxes next to all the items in the drop-down list if the field has been set to allow multiple values.

- An index is a list generated by Access containing pointers that direct Access to the locations of specific records in a table.

- Access automatically generates an index for a field designated the primary key field in a table.
- To create a single-field table index, change the Indexed property to *Yes (Duplicates OK)* or *Yes (No Duplicates)*.
- To create a multiple-field index, open the Indexes window.
- Normalizing a database involves reviewing the database structure to ensure that the tables are set up to eliminate redundancy. Three normalization states are tested: first normal form, second normal form, and third normal form.

Commands Review

FEATURE	RIBBON TAB, GROUP	BUTTON
Edit Relationships dialog box	Relationship Tools Design, Tools	
indexes window	Table Tools Design, Show/Hide	
primary key field	Table Tools Design, Tools	
relationships report	Relationship Tools Design, Tools	
Relationships window	Database Tools, Relationships OR Table Tools Design, Relationships	
Show Table dialog box	Relationship Tools Design, Relationships	

Microsoft®
Access®
Advanced Query Techniques

Performance Objectives

Upon successful completion of Chapter 3, you will be able to:

1 Save a filter as a query

2 Create and run a parameter query to prompt for criteria

3 Create an inner join, left outer join, and right outer join to modify query results

4 Add tables to and remove tables from a query

5 Create a self-join query to match two fields in the same table

6 Assign an alias to a table name

7 Create a query that includes a subquery

8 Create a query that uses conditional logic

9 Select records using a multiple-value field in a query

10 Create a new table using a make table query

11 Remove records from a table using a delete query

12 Add records to a table using an append query

13 Modify records using an update query

In this chapter, you will create, save, and run queries that incorporate advanced query features, such as saving a filter as a query, prompting for criteria on single and multiple fields, modifying join properties to view alternative query results, and using action queries to perform operations on groups of records. In addition, you will create an alias for a table and incorporate subqueries to manage multiple calculations.

 Data Files

Before beginning chapter work, copy the AL2C3 folder to your storage medium and then make AL2C3 the active folder.

The online course includes additional training and assessment resources.

You will create queries to select records by saving the criteria for a filter and creating a query that prompts the user for the criteria when the query is run.

Extracting Records Using Select Queries

A select query is the type of query most often used in Access. It extracts records that meet specified criteria from a single table or multiple tables. The subset of records that a query returns can be edited, viewed, and/or printed. In Query Design view, the criteria used to select records are entered by typing expressions in the *Criteria* row for the required field(s). Access also provides other methods for specifying query criteria.

Tutorial

Saving a Filter as a Query

Quick Steps

Save Filter as Query
1. Open table.
2. Filter table as needed.
3. Click Advanced button.
4. Click *Filter By Form*.
5. Click Advanced button.
6. Click *Save As Query*.
7. Type query name.
8. Click OK.

Advanced

Saving a Filter as a Query

A filter is used in a datasheet or form to temporarily hide records that do not meet specified criteria. For example, filter a WorkOrders datasheet to display only those work orders completed on a specified date. The subset of records can then be edited, viewed, and/or printed. Use the Filter by Form feature to filter a datasheet by multiple criteria using a blank datasheet. A filter is active until it is removed or the datasheet or form is closed. A filter can be saved when the table is closed but when the worksheet is reopened, all the records redisplay. Click the Toggle Filter button to reapply the filter.

If these steps are being repeated often, save the filtered datasheet as a query. Some users are more comfortable using filters to select records than with typing criteria expressions in Query Design view. To save a filter as a query, click the Advanced button in the Sort & Filter group on the Home tab and then click *Filter By Form* at the drop-down list to display the criteria. Click the Advanced button again and then click *Save As Query* at the drop-down list. Type a query name at the Save As Query dialog box and then press the Enter key or click OK. Saving a filter as a query means that all the columns in the table display in the query results datasheet. Use the Hide Fields feature to remove a field(s) from the results.

Activity 1a Saving a Filter as a Query   Part 1 of 3

1. Open **3-RSRCompServ** and enable the content.
2. Use the Filter by Form feature to display only those service calls that required two or more hours of labor by technicians billed at $50.00 per hour by completing the following steps:
   a. Open the WorkOrders table in Datasheet view.
   b. Minimize the Navigation pane.
   c. Hide the *Comments* field by right-clicking the *Comments* column heading in the datasheet and then clicking *Hide Fields* at the shortcut menu.

d. Click the Advanced button in the Sort & Filter group on the Home tab and then click *Filter By Form* at the drop-down list.

e. Click in the empty record in the *Hours* column. Type >=2 and then press the Tab key.

f. With the insertion point positioned in the *Rate* column, click the option box arrow that appears and then click *50* at the drop-down list.

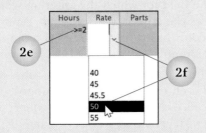

g. Click the Toggle Filter button (which displays the ScreenTip *Apply Filter*) in the Sort & Filter group on the Home tab. The records that meet the filter conditions display.

3. Review the seven filtered records in the datasheet.

4. Click the Toggle Filter button (which displays the ScreenTip *Remove Filter*) to redisplay all the records.

5. Click the Advanced button and then click *Filter By Form* at the drop-down list. Notice that the filter criteria in the *Hours* and *Rate* columns are intact.

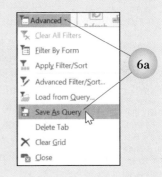

6. Save the filter as a query so the criteria can be reused by completing the following steps:

a. Click the Advanced button and then click *Save As Query* at the drop-down list.

b. At the Save As Query dialog box, type WO2orMoreRate50 in the *Query Name* text box and then click OK.

c. Click the Advanced button and then click *Close* at the drop-down list to close the Filter By Form datasheet.

d. Close the WorkOrders table. Click No when prompted to save changes to the table design.

7. Expand the Navigation pane.

8. Double-click the query object *WO2orMoreRate50* to open the query and then review the results.

9. Hide the *Comments* column in the query results datasheet.

10. Print the datasheet in landscape orientation with the left and right margins set to 0.5 inch.

11. Switch to Design view. Notice that the query design grid for a query created from a filter includes columns only for those columns for which criteria have been defined.

12. Close the query. Click Yes when prompted to save changes to the query layout.

Check Your Work

Tutorial

Creating a
Parameter Query

Creating a Parameter Query

In a parameter query, specific criteria for a field are not stored with the query design. Instead, the field(s) used to select records has a prompt message that displays when the query is run. The prompt message instructs the user to type the criteria to be used in selecting records.

Figure 3.1 shows the Enter Parameter Value dialog box, which displays when a parameter query to select by a technician's name is run. The message shown in the dialog box is created in the field for which the criterion will be applied. When the query is run, the user types the criterion at the Enter Parameter Value dialog box and Access selects the records based on the entry. If more than one field contains a parameter, Access prompts the user one field at a time.

A parameter query is useful if a query is run several times on the same field but different criteria are used each time. For example, suppose that a list of each technician's work orders is needed. Normally, a separate query would be made for each technician, resulting in several query objects being displayed in the Navigation pane. Creating a parameter query that prompts the user to enter the technician's name means that only one query is created.

To create a parameter query, start a new query in Design view and then add the desired tables and fields to the query design grid. Type a message enclosed in square brackets to prompt the user for the required criterion in the *Criteria* row of the field to be used to select records. Access does not allow punctuation at the end of the message. The text inside the square brackets is displayed in the Enter Parameter Value dialog box when the query is run. Figure 3.2 displays the entry in the *Criteria* row of the *FName* field that generated the Enter Parameter Value message shown in Figure 3.1.

Quick Steps

Create Parameter Query
1. Start new query in Design view.
2. Add table(s).
3. Close Show Table dialog box.
4. Add fields to query design grid.
5. Click in *Criteria* row in field to be prompted.
6. Type message text enclosed in square brackets.
7. Repeat Steps 5–6 for each additional criterion field.
8. Save and close query.

Hint If you are creating a parameter query that will be used by other people, consider adding an example of an acceptable entry in the message. For example, the message *Type the service date in the format mmm-dd-yyyy (example Oct-31-2018)* is more informative than *Type the service date*.

Hint Data typed in the Enter Parameter Value dialog box can be typed in lower case even if the field has upper case letters as Access is not case-sensitive when entering text strings.

Query Design

Figure 3.1 Enter Parameter Value Dialog Box

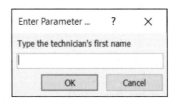

Figure 3.2 Criterion to Prompt for the Name in the *FName* Field

Type a message enclosed in square brackets to prompt the user for the criterion to use in selecting records.

Activity 1b Creating a Parameter Query to Prompt for Technician Names   Part 2 of 3

1. With **3-RSRCompServ** open, create a query in Design view to select records from the Technicians table and WorkOrders table by completing the following steps:
   a. Click the Create tab and then click the Query Design button in the Queries group.
   b. At the Show Table dialog box, add the Technicians table and WorkOrders table to the query.
   c. Close the Show Table dialog box.

d. At the top of the query, drag down the bottom borders of both table field list boxes until all the field names are visible. If necessary, resize the query design grid.

e. Double-click the following field names in the Technicians table field list box and WorkOrders table field list box to add the fields to the query design grid (click the field names in the order indicated): *WO, FName, LName, ServDate, Hours, Rate.*

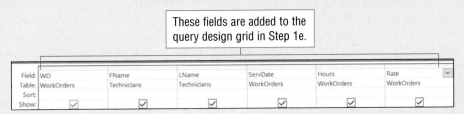

These fields are added to the query design grid in Step 1e.

2. Click the Run button in the Results group on the Query Tools Design tab to run the query.

3. Add parameters to select records by a technician's first and last names by completing the following steps:

a. Switch to Design view.

b. Click in the *Criteria* row in the *FName* column in the query design grid, type [Type the technician's first name], and then press the Enter key.

c. Position the mouse pointer on the vertical line between *FName* and *LName* in the gray field selector bar above the field names until the pointer changes to a left-and-right-pointing arrow with a vertical line in the middle. Double-click to expand the width of the *FName* column so the entire criterion entry can be seen.

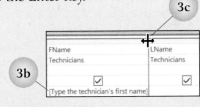

d. With the insertion point positioned in the *Criteria* row in the *LName* column, type [Type the technician's last name] and then press the Enter key.

e. Expand the width of the *LName* column so the entire criterion entry can be seen.

4. Click the Save button on the Quick Access Toolbar, type PromptedTechLabor in the *Query Name* text box at the Save As dialog box, and then click OK.

5. Close the query.

6. Run the parameter query and extract a list of work orders for the technician named *Pat Hynes* by completing the following steps:

a. Double-click the query named *PromptedTechLabor* in the Navigation pane.

b. Type pat at the first Enter Parameter Value dialog box, which displays the message *Type the technician's first name,* and then click OK. (Note that Access is not case-sensitive for text strings.)

c. Type hynes at the second Enter Parameter Value dialog box, which displays the message *Type the technician's last name,* and then click OK.

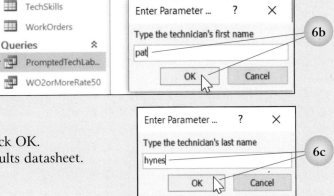

7. Review the records in the query results datasheet.

8. Print the query results datasheet.

9. Close the query.

Check Your Work

1. With **3-RSRCompServ** open, create a parameter query in Design view to prompt the user for the starting and ending dates for selecting records in the WorkOrders table by completing the following steps:
   a. Click the Create tab and then click the Query Design button.
   b. At the Show Table dialog box, add the WorkOrders table to the query and then close the dialog box.
   c. Drag down the bottom border of the table field list box until all the field names are visible. If necessary, resize the query design grid.
   d. Add the following fields to the query design grid in this order: *WO*, *CustID*, *Descr*, *ServDate*, *Hours*, *Rate*, *Parts*.
   e. Click in the *Criteria* row in the ServDate column, type the entry between [Type starting date] and [Type ending date], and then press the Enter key. Access capitalizes *Between* and *And*.

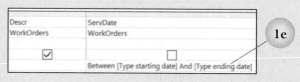

   f. Expand the width of the *ServDate* column until the entire criterion entry is visible.
2. Save the query, typing PromptedServiceDate as the query name, and then click OK.
3. Close the query.
4. Double-click *PromptedServiceDate* in the Navigation pane. At the first Enter Parameter Value dialog box with *Type starting date* displayed, type November 1, 2021 and then click OK. At the second Enter Parameter Value dialog box with *Type ending date* displayed, type November 13, 2021 and then click OK.

Work Order	Customer ID	Description	Service Date	Hours	Rate	Parts
65012	1000	Biannual computer maintenance	Mon, Nov 01 2021	1.25	50.00	$10.15
65013	1000	Replace keyboard	Wed, Nov 03 2021	0.50	45.50	$42.75
65014	1005	Replace power supply	Wed, Nov 03 2021	1.75	50.00	$62.77
65015	1008	Restore operating system	Fri, Nov 05 2021	2.25	50.00	$0.00
65016	1010	Install upgraded video card	Fri, Nov 05 2021	1.00	50.00	$48.75
65017	1015	Replace hard drive	Tue, Nov 09 2021	2.50	50.00	$55.87
65018	1020	Upgrade Office Suite	Thu, Nov 11 2021	1.00	45.00	$0.00
65019	1025	Upgrade to new Windows	Thu, Nov 11 2021	1.50	55.00	$0.00

Records within the date range November 1, 2021 to November 13, 2021 are selected in Step 4.

5. Print the query results datasheet in landscape orientation.
6. Close the query.

Check Your Work

Activity 2 **Modify Query Results by Changing the Join Property**   **4 Parts**

You will create and modify queries that obtain various results based on changing the join properties for related tables.

Modifying Join Properties in a Query

The term *join properties* refers to how Access matches the field values in the common field between the two tables in a relationship. The method used determines how many records are selected for inclusion in the query results datasheet. Access provides for three join types in a relationship: an inner join, a left outer join, and a right outer join. By default, Access uses an inner join between the tables; records are selected for display in a query only when a match on the joined field value exists in both tables. If a record exists in either table with no matching record in the other table, the record is not displayed in the query results datasheet. This means that in some cases, not all the records are seen from both tables when a query is run.

Hint Edit a relationship if the join type is always to be a left or right outer join. To do this, click the Join Type button in the Edit Relationships dialog box to open the Join Properties dialog box. Select the desired join type and then click OK.

Specifying the Join Type

Quick Steps
Create Query with Left Outer Join
1. Create new query in Design view.
2. Add tables to query window.
3. Double-click join line between tables.
4. Select option 2.
5. Click OK.
6. Add fields to query design grid.
7. Save and run query.

Double-click the join line between tables in a query to open the Join Properties dialog box, shown in Figure 3.3. By default, option *1*, referred to as an *inner join*, is selected. In an inner join, only those records are displayed for which the primary key field value in the primary table matches a foreign key field value in the related table.

Options *2* and *3* are referred to as *outer joins*. Option *2* is a left outer join. In this type of join, the primary table (referred to as the *left table*) displays all the rows, whereas the related table (referred to as the *right table*) displays only rows with matching values in the foreign key field. For example, examine the query results datasheet shown in Figure 3.4. This query was created with the Technicians table and TechSkills table. All the technician records are shown in the datasheet. However, notice that some technician records display with empty field values for the columns from the TechSkills table. These records reflect the technicians for which no information has yet been entered in the TechSkills table. In a left outer join, all the records from the primary table in the relationship are shown in the query results datasheet.

Hint Changing the join type at a query window does not change the join type for other objects based on the relationship. The revised join property applies only to the query.

Figure 3.3 Join Properties Dialog Box

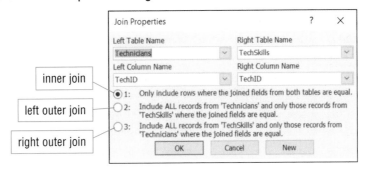

Figure 3.4 Left Outer Join Example

First Name	Last Name	Certifications	Operating Systems	Network Specialist?	Design Websites?	Programming?
Pat	Hynes	CCNA Cloud, CCNA Wireless, Microsoft MCT	Linux, Unix, Windows 7, Windows 8	✓		✓
Hui	Chen	CCNA Cloud, Comp TIA A+	Linux, Unix, Windows 10, Windows 7, Windo	✓	✓	
Kayla	Blewett			■	■	■
Mihir	Patel	Comp TIA A+, Microsoft MCT	Unix, Windows 10, Windows 8			✓
Madir	Sadiku	Comp TIA A+, Microsoft MCT	Mac OS X, Windows 10, Windows 7, Windov			
Brody	Stewart			■	■	■
Ana	Takacs	CCNA Wireless	Windows 8	✓		✓
Marty	O'Leary	CCNA Wireless	Linux, Unix	✓	✓	
Armando	Colacci	CCNA Wireless, MOS Master	Windows 10, Windows 7, Windows 8			✓
Kelsey	Eastman	Comp TIA A+, MOS Master	Linux, Windows 7, Windows 8	✓	✓	
Dana	Westman			■	■	■
				■	■	■

Left outer join query results show blank related *TechSkills* fields for those technicians for which records have not yet been entered in the TechSkills table.

Quick Steps

Create Query with Right Outer Join
1. Create new query in Design view.
2. Add tables to query window.
3. Double-click join line between tables.
4. Select option *3*.
5. Click OK.
6. Add fields to query design grid.
7. Save and run query.

Option *3* is a right outer join. In this type of join, the related table (right table) shows all the rows, whereas the primary table (left table) shows only rows with matching values in the common field. For example, examine the partial query results datasheet shown in Figure 3.5. This datasheet illustrates 15 of the 39 records in the query results datasheet from the Technicians table and WorkOrders table. Notice that the first four records have no technician first or last names. These are the work orders that have not yet been assigned to a technician. In a right outer join, all the records from the related table in the relationship are shown in the query results datasheet. In a left or right outer join, Access displays an arrow at the end of the join line pointing to the table that shows only matching rows. To display only the records that have not yet been assigned to a technician, type *null* in the criteria of either the first or last name fields.

To illustrate the difference in query results when no change is made to the join type, examine the query results datasheet shown in Figure 3.6. This is the datasheet created in Activity 2a. In this activity, a list of technician names and qualifications will be created. Compare the number of records shown in Figure 3.6 with the number of records shown in Figure 3.4. Notice that fewer records display in the datasheet in Figure 3.6. Since an inner join displays only those records for which matching entries exists in both tables, records from either table that do not have matching records in the other table are not displayed. Understanding that an inner join (the default join type) may not display all the records that exist in the tables when a query is run is important.

Figure 3.5 Right Outer Join Example

Right outer join query results show blank related technician fields for those work orders that have not yet been assigned to a technician.

First Name	Last Name	Work Order	WO Date	Description
		65047	Mon, Nov 29 2021	Set up automatic backup
		65048	Mon, Nov 29 2021	Replace LCD monitor
		65049	Tue, Nov 30 2021	Set up dual monitor system
		65050	Tue, Nov 30 2021	Reinstall Windows 10
Pat	Hynes	65020	Fri, Nov 12 2021	Troubleshoot noisy fan
Pat	Hynes	65033	Tue, Nov 23 2021	Install Windows 10 and Office 365
Pat	Hynes	65038	Fri, Nov 26 2021	Install latest version of Windows
Hui	Chen	65014	Wed, Nov 03 2021	Replace power supply
Hui	Chen	65019	Tue, Nov 09 2021	Upgrade to new Windows
Hui	Chen	65026	Wed, Nov 17 2021	Upgrade RAM
Hui	Chen	65032	Tue, Nov 23 2021	Install second storage drive
Hui	Chen	65035	Tue, Nov 23 2021	Office 365 training
Kayla	Blewett	65023	Mon, Nov 15 2021	Upgrade RAM
Kayla	Blewett	65036	Wed, Nov 24 2021	Set up home network
Kayla	Blewett	65041	Fri, Nov 26 2021	Biannual computer maintenance

Figure 3.6 Inner Join Example

First Name	Last Name	Certifications	Operating Systems	Network Specialist?	Design Websites?	Programming?
Pat	Hynes	CCNA Cloud, CCNA Wireless, Microsoft MCT	Linux, Unix, Windows 7, Windows 8	✓		✓
Hui	Chen	CCNA Cloud, Comp TIA A+	Linux, Unix, Windows 10, Windows 7, Windc	✓	✓	
Mihir	Patel	Comp TIA A+, Microsoft MCT	Unix, Windows 10, Windows 8			✓
Madir	Sadiku	Comp TIA A+, Microsoft MCT	Mac OS X, Windows 10, Windows 7, Windov			
Ana	Takacs	CCNA Wireless	Windows 8	✓		✓
Marty	O'Leary	CCNA Wireless	Linux, Unix	✓	✓	
Armando	Colacci	CCNA Wireless, MOS Master	Windows 10, Windows 7, Windows 8	✓		✓
Kelsey	Eastman	Comp TIA A+, MOS Master	Linux, Windows 7, Windows 8	✓	✓	

An inner join displays a record from either table only when a matching value in the joined field exists in the other table. No blank records appear in the query results. However, notice that not all the records in the Technicians table display.

Activity 2a Selecting Records in a Query Using an Inner Join   Part 1 of 4

1. With **3-RSRCompServ** open, create a query in Design view to display a list of technicians that notes each individual's skill specialties by completing the following steps:
   a. Create a new query in Design view. At the Show Table dialog box, add the Technicians table and TechSkills table. Close the Show Table dialog box and then drag down the bottom borders of both table field list boxes until all the field names are visible. If necessary, resize the query design grid.
   b. Double-click the join line between the two tables to open the Join Properties dialog box.

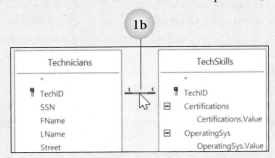

   c. At the Join Properties dialog box, notice that option *1* is selected by default. Option *1* selects records only when the joined fields from both tables are equal. This represents an inner join. Click OK.

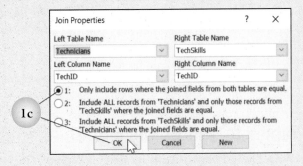

   d. Add the following fields to the query design grid in order: *FName, LName, Certifications, OperatingSys, NetworkSpc, WebDesign, Programming*.
   e. Run the query.
2. Save the query, typing TechSpecialties as the query name, and then click OK.
3. Close the query.

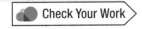

1. With **3-RSRCompServ** open, modify the TechSpecialties query to a left outer join to check whether information for any technicians has not yet been entered in the TechSkills table by completing the following steps:
   a. Right-click the *TechSpecialties* query and then click *Design View* at the shortcut menu.
   b. Right-click the join line between the two tables and then click *Join Properties* at the shortcut menu.

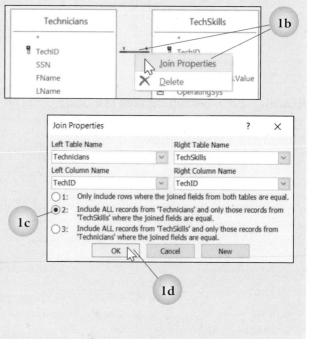

   c. At the Join Properties dialog box, click option *2*. Option *2* includes all the records from the Technicians table and only those records from the TechSkills table for which the joined fields are equal. The left table (Technicians) is the table that will show all the records. If there is not a matching record for a technician in the other table, the columns display empty fields next to the technician's name. *Note: Do not assume that a left join always occurs with the table that is the left table in the query window. Although Technicians is the left table in the query window, the term "left" refers to the table that represents the "one" side (primary table) in the relationship.*
   d. Click OK.
   e. Notice that the join line between the two tables now displays with an arrow pointing to the joined field in the TechSkills table.
   f. Run the query.
2. Minimize the Navigation pane and then compare your results with the query results datasheet displayed in Figure 3.4 (on page 60). Notice that 11 records display in this datasheet, whereas only 8 records display in the query results from Activity 2a.
3. Click the File tab and then click the *Save As* option. At the Save As backstage area, click *Save Object As* in the *File Types* section and then click the Save As button. At the Save As dialog box, type AllTechSkills in the *Save 'TechSpecialties' to* text box and then click OK.
4. Close the query.
5. Expand the Navigation pane.

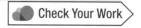

Adding Tables to and Removing Tables from a Query

Show Table — Open a query in Design view to add a table to a query. Click the Show Table button in the Query Setup group on the Query Tools Design tab and then add the desired table using the Show Table dialog box. Close the Show Table dialog box when finished.

To remove a table from a query, click any field within the table field list box to activate the table in the query window and then press the Delete key. The table is removed from the window and all the fields associated with the table that were added to the query design grid are automatically removed. A table can also be removed by right-clicking the table in the query window and then clicking *Remove Table* at the shortcut menu.

Activity 2c Selecting Records in a Query Using a Right Outer Join   Part 3 of 4

1. With **3-RSRCompServ** open, modify an existing query to create a new query to check for work orders that have not been assigned to a technician by completing the following steps:
   a. Open the TechSpecialties query in Design view.
   b. Right-click the *TechSkills* table name in the table field list box in the query window and then click *Remove Table* at the shortcut menu. Notice that the last five columns are removed from the query design grid along with the table.

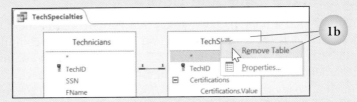

   c. Click the Show Table button in the Query Setup group on the Query Tools Design tab.
   d. At the Show Table dialog box, double-click *WorkOrders*, click the Close button, and then drag down the bottom border of the WorkOrders table field list box until all the fields are visible. If necessary, resize the query design grid.
   e. Double-click the join line between the two tables.
   f. At the Join Properties dialog box, click option *3*. Option *3* includes all the records from the WorkOrders table and only those records from the Technicians table for which the joined fields are equal. The right table (WorkOrders) is the table that will show all the records. If a work order does not have a matching record in the other table, the columns display empty fields for the technician names. ***Note: Do not assume that a right join always occurs with the table that is the right table in the query window. Although WorkOrders is the right table in the query window, the term "right" refers to the table that represents the "many" side (related table) in the relationship***.

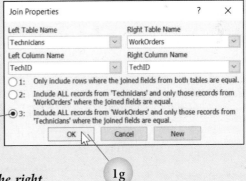

   g. Click OK.
   h. Notice that the join line between the two tables now displays with an arrow pointing to the joined field in the Technicians table.
   i. Add the following fields from the WorkOrders table to the query design grid in this order: *WO, WODate, Descr*.
2. Click the File tab and then click the *Save As* option. At the Save As backstage area, click *Save Object As* in the *File Types* section and then click the Save As button. At the Save As dialog box, type UnassignedWO in the *Save 'TechSpecialties' to* text box and then click OK.
3. Run the query.

4. Compare your results with the partial query results datasheet shown in Figure 3.5 (on page 60). Notice that the first four records in the query results datasheet have empty fields in the *First Name* column and *Last Name* column.
5. Double-click the right boundary of the *Description* column to adjust the width.
6. Print only the records that have not been assigned to a technician by completing the following steps.
   a. Switch to Design view.
   b. Click in the *FName Criteria* row and then type null.
   c. Run the query.
   d. Print the query results datasheet with the right margin set to 0.5 inch.
7. Close the query. Click Yes to save the changes.

 Tutorial

Creating a Self-Join Query and Assigning Aliases

💡 **Hint** In a self-join query, the two fields joined must have the same data type.

Quick Steps

Create Self-Join Query and Assign Aliases

1. Create query in Design view.
2. Add two copies of table to query.
3. Right-click second table name.
4. Click *Properties*.
5. Click in *Alias* property box and delete existing table name.
6. Type alias table name.
7. Close Property Sheet.
8. Drag field name from left table to field name with matching values in right table.
9. Add fields to query design grid.
10. Save and run query.

Creating a Self-Join Query

Assume that a table has two fields that contain similar field values. For example, look at the *Technician ID* column and *Tier 2 Supervisor* column in the Technicians table datasheet, shown in Figure 3.7. Notice that each column contains a technician's ID number. Tier 2 supervisors are senior technicians who are called in when a work order is too complex for a regular technician to solve. The ID number in the *Tier 2 Supervisor* column is the ID of the senior technician who is assigned to the technician.

Viewing the list of technicians with the Tier 2 supervisors' last names may be more informative than viewing the list with the supervisors' ID numbers. If a table has matching values in two separate fields, create a self-join query; this creates a relationship between fields in the same table. To create a self-join query, add two copies of the same table to the query window. The second occurrence of the table is named using the original table name with *_1* added to the end. Assign an alias to the second table to provide it with a more descriptive name in the query. To join the two tables, drag the field with matching values from one table field list to the other. Add the required fields to the query design grid and then run the query.

Creating an Alias for a Table

An alias is an additional name that can be used to reference a table in a query. The alias is temporary and applies only to the query. Generally, the reason for creating an alias is to assign a shorter name to a table (or a more descriptive name, in the case of a self-join query). For example, one of the tables used in the query in Activity 2d is named *Technicians_1*. Assign the table a more descriptive name, such as *Supervisors*, to more accurately describe its role in the query.

To assign an alias to a table, right-click the table name in the query window and then click *Properties* at the shortcut menu to open the Property Sheet task pane. Click in the *Alias* property box, delete the existing table name, and then type the reference name of the table. Access replaces all the occurrences of the table name in the query design grid with the alias.

Figure 3.7 Technicians Table Datasheet with the Fields Used in a Self-Join Query

Technician ID	SSN	First Name	Last Name	Street Address	City	State	ZIP Code	Home Phone	Cell Phone	🔗	Tier 2 Supervisor
⊞ 01	000-45-5368	Pat	Hynes	206-31 Woodland Street	Detroit	MI	48202-1138	313-555-6874	313-555-6412	⑴	03
⊞ 02	000-47-3258	Hui	Chen	12905 Hickory Street	Detroit	MI	48205-3462	313-555-7468	313-555-5234	⑴	06
⊞ 03	000-62-7468	Kayla	Blewett	1310 Jarvis Street	Detroit	MI	48220-2011	313-555-3265	313-555-6486	⑴	
⊞ 04	000-33-1485	Mihir	Patel	8213 Elgin Street	Detroit	MI	48234-4092	313-555-7458	313-555-6385	⑴	11
⊞ 05	000-48-7850	Madir	Sadiku	8190 Kenwood Street	Detroit	MI	48220-1132	313-555-6327	313-555-8569	⑴	03
⊞ 06	000-75-8412	Brody	Stewart	3522 Moore Place	Detroit	MI	48208-1032	313-555-7499	313-555-3625	⑴	
⊞ 07	000-55-1248	Ana	Takacs	14902 Hampton Court	Detroit	MI	48215-3616	313-555-6142	313-555-4586	⑼	11
⊞ 08	000-63-1247	Marty	O'Leary	14000 Vernon Drive	Detroit	MI	48237-1320	313-555-9856	313-555-4125	⑼	11
⊞ 09	000-84-1254	Armando	Colacci	17302 Windsor Avenue	Detroit	MI	48224-2257	313-555-9641	313-555-8796	⑼	06
⊞ 10	000-43-5789	Kelsey	Eastman	550 Montclair Street	Detroit	MI	48214-3274	313-555-6315	313-555-7411	⑵	06
⊞ 11	000-65-4185	Dana	Westman	18101 Keeler Streeet	Detroit	MI	48223-1322	313-555-5488	313-555-4158	⑼	
✳										⑼	

> These two fields contain technician ID numbers. Tier 2 supervisors are senior-level technicians who are assigned to handle complex cases for regular technicians.

Activity 2d Creating a Self-Join Query   Part 4 of 4

1. With **3-RSRCompServ** open, create a self-join query to display the last name of the Tier 2 supervisor instead of his or her ID number by completing the following steps:

   a. Create a new query in Design view.

   b. At the Show Table dialog box, double-click *Technicians* two times to add two copies of the Technicians table to the query and then close the Show Table dialog box. Notice that the second copy of the table is named *Technicians_1*.

   c. Drag down the bottom borders of both table field list boxes until all the field names are visible. If necessary, resize the query design grid.

   d. Create an alias for the second table by completing the following steps:

      1) Right-click the *Technicians_1* table name and then click *Properties* at the shortcut menu.

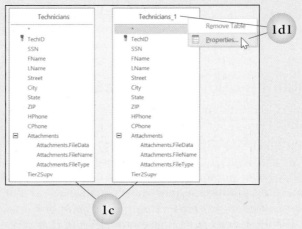

      2) Select the current name in the *Alias* property box in the Property Sheet task pane and then type Supervisors.

      3) Close the Property Sheet task pane.

e. Drag the field named *Tier2Supv* from the Technicians table field list box at the left to the field named *TechID* in the Supervisors table field list box at the right. This creates a join line between the two tables.

f. Add the *FName* and *LName* fields from the Technicians table to the query design grid.

g. Add the *LName* field from the Supervisors table to the query design grid.

2. Run the query. The last names displayed in the second *Last Name* column represent the *Tier2Supv* names.

3. Switch to Design view.

4. Right-click the second *LName* in the *Field* row (from the Supervisors table) in the query design grid and then click *Properties* at the shortcut menu. Click in the *Caption* property box in the Property Sheet task pane, type Tier 2 Supervisor, and then close the Property Sheet task pane.

5. Save the query, typing Tier2Supervisors as the query name, and then click OK.

6. Run the query.

7. Double-click the right boundary of the *Tier 2 Supervisor* column to adjust the width and then compare your results with the datasheet shown at the right.

8. Close the query. Click Yes to save changes.

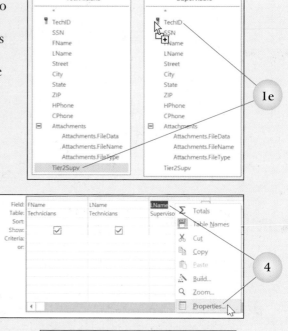

Check Your Work

Running a Query with No Established Relationship

If a query is created from two tables for which no join has been established, Access will not know how to relate the records in the tables. When there is no relationship, Access produces a datasheet representing every possible combination of records between the two tables. For example, if one table contains 20 records, the other table contains 10 records, and no join has been established between the tables, Access produces a query results datasheet containing 200 records (20×10). This type of query is called a *cross product query* or *Cartesian product query*. In most cases, the results of such a query provide data that serves no purpose.

If two tables are added to a query and no join line appears, create a join by dragging a field from one table field list box to a compatible field in the other table field list box. The two fields should contain the same data type and be logically related in some way. If no join can logically be established and a many-to-many relationship exists between the tables, add a junction table to the query.

You will use a subquery nested within another query to calculate the total amount earned from each work order. You will also use conditional logic to apply a discount if certain criteria are met.

Creating and Using
Subqueries

Review: Performing
Calcuations in a
Query

Q̇uick Steps

Create Subquery
1. Start new query in
   Design view.
2. At Show Table dialog
   box, click Queries
   tab.
3. Double-click query to
   be used as subquery.
4. Add other queries or
   tables as required.
5. Close Show Table
   dialog box.
6. Add fields as
   required.
7. Save and run query.

 Builder

Creating and Using Subqueries

When performing multiple calculations based on numeric fields, a user may decide to create a separate query for each individual calculation and then use subqueries to generate the final total. The term *subquery* is used to refer to a query nested inside another query. Using subqueries to break calculations into individual objects allows a calculated field to be reused in multiple queries.

For example, assume that the total amount for each work order needs to be calculated. The WorkOrders table contains fields with the number of hours for each service call, the labor rate, and the total value of the parts used. To find the total for each work order, the total labor needs to be calculated. This is done by multiplying the hours times the rate and then adding the parts value to the total labor value. However, the total labor value should be in a separate query so that other calculations can be performed, such as finding the average, maximum, and minimum labor on work orders. To be able to reuse the total labor value, create the calculated field in its own query.

Level 1, Chapter 3 demonstrated how to insert a calculated field in a query. Recall that the format for inserting an equation in a query is to type in a blank *Field* row the new field name followed by a colon and then the equation with the field names in square brackets—for example, *Total:[Sales]+[SalesTax]*. If the equation is typed in the Expression Builder, the square brackets can be omitted from the field names as Access will add them automatically. If an Enter Parameter dialog box displays when the query is run, check the spelling of the field names. Also check to make sure that the correct type of brackets have been used and that there are no extra spaces.

Activity 3a **Creating a Query to Calculate Total Labor**   **Part 1 of 3**

1. With **3-RSRCompServ** open, create a query to calculate the total labor for each work order by completing the following steps:
   a. Create a new query in Design view. At the Show Table dialog box, add the WorkOrders table to the query window and then close the Show Table dialog box.
   b. Drag down the bottom border of the WorkOrders table field list box until all the fields are visible. If necessary, resize the query design grid.
   c. Add the following fields to the query design grid in this order: *WO, ServDate, Hours, Rate*.
   d. Click in the blank *Field* row next to *Rate* in the query design grid and then click the Builder button in the Query Setup group on the Query Tools Design tab.
   e. In the Expression Builder dialog box, type TotalLabor:Hours*Rate.
   f. Click OK.
   g. Expand the width of the calculated column until the entire formula in the *Field* row can be seen.

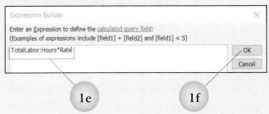

2. Run the query and view the query results. Notice that the *TotalLabor* column does not display a consistent number of decimal values.

3. Switch to Design view.

4. Format the *Total Labor* column by completing the following steps:

   a. Activate the field by clicking in the *TotalLabor* field row.

   b. Click the Property Sheet button in the Show/Hide group on the Query Tools Design tab.

   c. Click in the *Format* property box in the Property Sheet task pane, click the option box arrow that appears, and then click *Standard* at the drop-down list.

   d. Type Total Labor in the *Caption* property box.

   e. Close the Property Sheet task pane.

5. Save the query, typing TotalLabor as the query name, and then click OK.

6. Run the query. Notice that the last four rows contain no values because the service calls have not yet been completed.

7. Switch to Design view. Click in the *Criteria* row in the *Hours* column, type >0, and then press the Enter key.

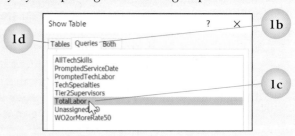

8. Save the revised query and then run it. Adjust any column widths if necessary.

9. Close the query.

> **Check Your Work**

> 💡 **Hint** Subqueries are not restricted to use in nested calculations. Use a subquery for any combination of fields that you want to reuse in multiple queries.

Once the query is created to calculate the total labor, nest the query inside another query to add the labor to the parts and then calculate the total for each work order. Creating subqueries provides flexibility in reusing calculations, thus avoiding duplication of effort and reducing the potential for calculation errors.

Small queries and subqueries are useful because they are easier to build and troubleshoot than large queries. Subqueries are also useful when creating a complex query. They can be created to build and test sections individually and then combined into the final larger query.

Activity 3b Creating a Subquery   Part 2 of 3

1. With **3-RSRCompServ** open, create a new query to calculate the total value of each work order using the TotalLabor query as a subquery by completing the following steps:

   a. Create a new query in Design view.

   b. At the Show Table dialog box, click the Queries tab.

   c. Double-click *TotalLabor* in the queries list box.

   d. Click the Tables tab.

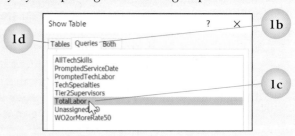

e. Double-click *WorkOrders* and then close the Show Table dialog box. Notice that Access automatically joins the two objects at the *WO* field.

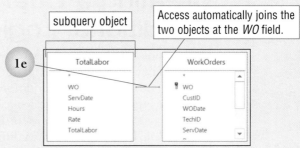

subquery object

Access automatically joins the two objects at the *WO* field.

2. Add fields from the TotalLabor subquery and WorkOrders table by completing the following steps:

a. Double-click the asterisk (*) at the top of the TotalLabor table field list box. Access adds the entry *TotalLabor.** to the first column in the query design grid. This entry adds all the fields from the query. Individual columns do not display in the grid but when the query is run, the datasheet will show all the fields.

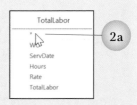

b. Run the query. Notice that the query results datasheet shows all five columns from the TotalLabor query.

c. Switch to Design view. Apply the Currency format to the *Total Labor* column in this new query. To do this, add the column to the query design grid as an individual field and not as a group by completing the following steps:

   1) Right-click in the field selector bar (the gray bar above the *Field* row) for the *TotalLabor.** column and then click *Cut* at the shortcut menu to remove the column from the query design grid.

   2) Add the following fields from the TotalLabor query field list box to the query design grid in this order: *WO, ServDate, TotalLabor*.

   3) Format the *TotalLabor* column by applying the Currency format.

d. Drag down the bottom border of the WorkOrders table field list box until all the fields are visible and then double-click *Parts* to add the field to the query design grid. If necessary, resize the query design grid.

3. Create a calculated field to add the total labor and parts by completing the following steps:

a. Click in the blank *Field* row next to *Parts* in the query design grid, type TotalWorkOrder:[TotalLabor]+[Parts], and then press the Enter key.

b. Expand the width of the *TotalWorkOrder* column until the entire formula is visible.

c. Click in the *TotalWorkOrder* column, open the Property Sheet task pane and type Total Work Order in the *Caption* property box.

e. Close the Property Sheet task pane.

4. Save the query, typing TotalWorkOrders as the query name, and then click OK. Run the query.

5. Double-click the right boundary for the *TotalWorkOrder* column to adjust the width.

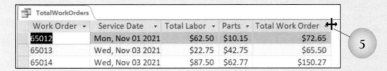

6. Close the query. Click Yes to save changes.

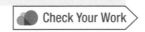
Check Your Work

Tutorial

Creating a Query
Using Conditional
Logic

Creating a Query Using Conditional Logic

Using conditional logic in an Access query is similar to using a logical formula in Excel. Using conditional logic requires Access to perform a calculation based on the outcome of a logical or conditional test. One calculation is performed if the test proves true and another calculation is performed if the test proves false. The structure of the IF function is *=IIF(logical_test,value_if_true,value_if_false)*. Access uses an *Immediate IF* (IIF) function to differentiate between this function and the Visual Basic for Applications (VBA) IF function. VBA is discussed further in Chapter 7. Logical functions are also similar to calculated fields in Access in that the new field name is followed by a colon and field names are enclosed in square brackets.

In Activity 3c, a 2% discount will be processed on the total cost of the work order if the cost of the labor is greater than $80. The following IF function will be used to calculate this discount, where *Labor>80* in the field that will store the discounted total: *Labor>80:IIF([TotalLabor]>80,[TotalWorkOrder]-([TotalWorkOrder]*.02),[TotalWorkOrder])*.

To determine whether a discount should be provided, Access first tests the value of the record in the *TotalLabor* field to see if it is greater than $80. If the condition proves true, Access populates the *Labor>80* field with the value from the *TotalWorkOrder* field minus 2%. This is the discounted total that the customer owes. If the value in the *TotalLabor* field is not greater than $80, the condition proves false. Access populates the *Labor>80* field with the same value that appears in the *TotalWorkOrder* field since the work order did not qualify for a discount.

Figure 3.8 shows the TotalWorkOrders query after the IF function has been applied. For work order 65019, the value in the *TotalLabor* field is $82.50. Since this value is greater than $80, the test is true and Access returns the value_if_true value, or the value in the *TotalWorkOrder* field minus 2% of the value ($82.50 – $1.65 = $80.85). For work order 65012, the value in the *TotalLabor* field is $72.65. Since the test is false, Access returns the value_if_false value, or the value in the *TotalWorkOrder* field ($72.65).

Figure 3.8 Activity 3c TotalWorkOrders Query in Datasheet View

Work Order	Service Date	Total Labor	Parts	Total Work Order	Work Order w/Disc
65012	Mon, Nov 01 2021	$62.50	$10.15	$72.65	$72.65
65013	Wed, Nov 03 2021	$22.75	$42.75	$65.50	$65.50
65014	Wed, Nov 03 2021	$87.50	$62.77	$150.27	$147.26
65015	Fri, Nov 05 2021	$112.50	$0.00	$112.50	$110.25
65016	Fri, Nov 05 2021	$50.00	$48.75	$98.75	$98.75
65017	Tue, Nov 09 2021	$125.00	$55.87	$180.87	$177.25
65018	Thu, Nov 11 2021	$45.00	$0.00	$45.00	$45.00
65019	Thu, Nov 11 2021	$82.50	$0.00	$82.50	$80.85
65020	Mon, Nov 15 2021	$75.00	$72.50	$147.50	$147.50
65021	Mon, Nov 15 2021	$178.75	$0.00	$178.75	$175.18
65022	Tue, Nov 16 2021	$82.50	$400.00	$482.50	$472.85
65023	Mon, Nov 15 2021	$62.50	$100.00	$162.50	$162.50

The caption *Work Order w/Disc* was added to the *Labor>80* field.

Since the *Total Labor* value is less than $80, Access returns the value from the *Total Work Order* field in the *Work Order w/Disc* field.

Since the *Total Labor* value is greater than $80, Access subtracts 2% from the *Total Work Order* field and displays this value in the *Work Order w/Disc* field.

1. With **3-RSRCompServ** open, open the TotalWorkOrders query in Design view.
2. Modify the TotalWorkOrders query to calculate a 2% discount if the total labor cost is greater than $80 by completing the following steps:
   a. Right-click in the blank *Field* row next to *TotalWorkOrder* in the query design grid and then click *Zoom* at the drop-down list.

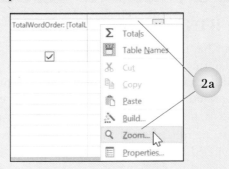

   b. Type Labor>80:IIF([TotalLabor]>80,[TotalWorkOrder]-([TotalWorkOrder]*.02), [TotalWorkOrder]).
   c. Click OK.

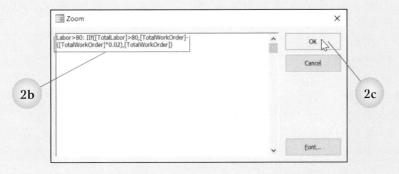

   d. Run the query. If an error message indicating that invalid syntax was entered or an Enter Parameter Value dialog box displays, close the message box and check to ensure that the formula was typed correctly.
3. Switch to Design view, open the Property Sheet task pane for the *Labor>80* column, and then make the following changes:
   a. Apply the Currency format.
   b. Type Work Order w/Disc in the *Caption* property box.
   c. Close the Property Sheet task pane.
4. Run the query.
5. Double-click the right boundary of the *Work Order w/Disc* column to adjust the width. Compare your results with the query results data displayed in Figure 3.8.
6. Print the query results datasheet with the right margin set to 0.5 inch.
7. Close the query. Click Yes to save changes.

Check Your Work

Activity 4 Query a Multiple-Value Field   **1 Part**

You will select records using a multiple-value field in a query.

Tutorial

Using a Multiple-Value Field in a Query

Quick Steps

Show Multiple-Value Field in Separate Rows in Query

1. Open query in Design view.
2. Click in *Field* row of multiple-value field in design grid.
3. Move insertion point to end of field name.
4. Type period (.).
5. Press Enter key to accept *.Value*.
6. Save query.

Using a Multiple-Value Field in a Query

Recall from Chapter 2 that a multiple-value field displays a lookup list that allows for more than one check box to be checked. In a query, a multiple-value field can display as it does in a table datasheet, with the multiple field values in the same column separated by commas. Or each field value can be shown in a separate row.

To show each value in a separate row, add *.Value* at the end of the multiple-value field name in the *Field* row in the query design grid. Figure 3.9 displays the query design grid for the query used in Activity 4 that displays each entry in the *OperatingSys* field in a separate row in the datasheet.

To select records using criteria in a multiple-value field, type the criteria using the same procedures as for a single-value field. For example, in the TechSpecialties query, typing *Windows 10* in the *Criteria* row in the *OperatingSys* column causes Access to return the records of any technician with Windows 10 certification as one of the multiple field values.

Figure 3.9 Activity 4 Query Design Grid

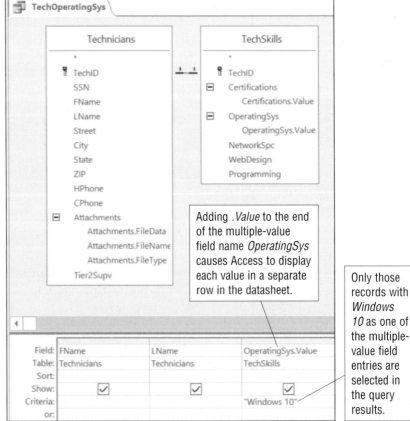

1. With **3-RSRCompServ** open, open the TechSpecialties query in Design view.
2. Right-click in the field selector bar above the *Certifications* field and then click *Cut* at the shortcut menu to remove the field from the query design grid.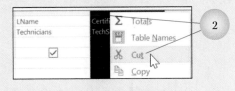
3. Delete the *NetworkSpc*, *WebDesign*, and *Programming* columns from the query design grid. Refer to Step 2.
4. Run the query. Notice that each record in the *Operating Systems* column displays the multiple values separated by commas.
5. Switch to Design view.
6. Click in the *OperatingSys* field row in the query design grid, move the insertion point to the end of the field name, and then type a period (.). Access automatically adds *.Value* to the end of the name in the *Field* row. Press the Enter key to accept *.Value* at the end of the field name. **Note: When creating a query from scratch, drag the multiple-value field name with the .Value property already attached from the table field list box to the query design grid.**
7. Save the query with a new name, typing TechOperatingSys as the query name, and then click OK.
8. Run the query. Notice that each entry in the multiple-value field now displays in a separate row.
9. Switch to Design view.
10. Click in the *Criteria* row in the *OperatingSys.Value* column in the query design grid, type Windows 10, and then press the Enter key. Access adds quotation marks around the text.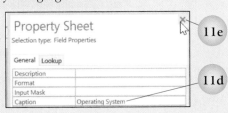
11. Run the query. Notice that the column title for the multiple-value field in the query results datasheet is now *TechSkills.OperatingSys. Value*. Change the column heading for the field by completing the following steps:
    a. Switch to Design view.
    b. Click in the *OperatingSys.Value* field row in the query design grid.
    c. Press the F4 function key to open the Property Sheet task pane.
    d. Click in the *Caption* property box, type Operating System, and then press the Enter key.
    e. Close the Property Sheet task pane.
12. Run the query.
13. Print the query results datasheet and then close the query. Click Yes to save changes.

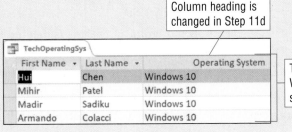

Column heading is changed in Step 11d

Technicians with Windows 10 are selected in Step 10.

Check Your Work

Activity 5 **Modify Records Using Action Queries**   **4 Parts**

You will create a new table, add and delete records, and update field values using action queries.

Performing Operations Using Action Queries

An action query is used to perform an operation on a group of records. Building an action query is similar to building a select query but with the extra step of specifying the action to perform on the group of selected records. Four types of action queries are available, as described in Table 3.1.

Hint Create a backup copy of the database before running an action query.

To create an action query, first build a select query by adding tables, fields, and criteria to the query design grid. Run the select query to make sure the correct group of records is targeted for action. After verifying the results that the proper records will be modified, change the query type using the Make Table, Append, Update, or Delete button in the Query Type group on the Query Tools Design tab, shown in Figure 3.10. Once the query type has been changed to an action query, clicking the Run button causes Access to perform the make table, append, update, or delete operation. Once an action query has been run, the results cannot be reversed.

When you build queries, Access creates code behind the scenes using Structured Query Language (SQL). View the code by clicking the View Button arrow and then clicking *SQL View*. Three queries, the Union, Pass-Through, and Data Definition queries, are created directly in the SQL view and can be initiated using buttons on the Query Tools Design tab in the Query Type group as shown in Figure 3.10.

Table 3.1 Types of Action Queries

Query Type	Description
make table	A new table is created from selected records in an existing table—for example, a new table that combines fields from two other tables in the database.
append	Selected records are added to the end of an existing table. This action is similar to performing a copy and paste.
update	A global change is made to the selected group of records based on an update expression—for example, increasing the labor rate by 10% in one step.
delete	The selected group of records is deleted from a table.

Figure 3.10 Query Type Group on the Query Tools Design Tab

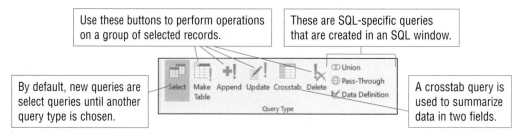

Use these buttons to perform operations on a group of selected records.

These are SQL-specific queries that are created in an SQL window.

By default, new queries are select queries until another query type is chosen.

A crosstab query is used to summarize data in two fields.

Select Make Table Append Update Crosstab Delete Union Pass-Through Data Definition

Query Type

Creating a New Table Using a Query

 Make Table

A make table query creates a new table from selected records in the same database or another database. This type of query is useful for creating a history table before purging old records that are no longer required. The history table can be placed in the same database or in another database used as an archive copy.

Once a select query is created that will extract the records to be copied to a new table, click the Make Table button in the Query Type group on the Query Tools Design tab. This opens the Make Table dialog box, shown in Figure 3.11. Enter a table name, choose the destination database, and then click OK. Once a make table query has been run, do not double-click the query name in the Navigation pane; doing so instructs Access to run the query again. Open the query in Design view to make changes to the criteria and/or query type if the query needs to be run again.

Quick Steps

Create Make Table Query
1. Create query in Design view.
2. Add table(s) to query.
3. Add fields to query design grid.
4. If necessary, enter criteria to select records.
5. Run query.
6. Switch to Design view.
7. Click Make Table button.
8. Type table name.
9. If necessary, select destination database.
10. Click OK.
11. Run query.
12. Click Yes.
13. Save query.

Figure 3.11 Make Table Dialog Box

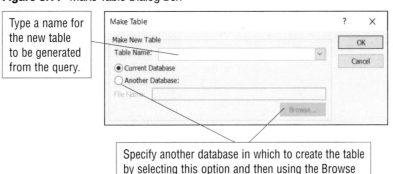

Type a name for the new table to be generated from the query.

Specify another database in which to create the table by selecting this option and then using the Browse button to navigate to the other database file name.

Activity 5a Creating a New Table Using a Query

Part 1 of 4

1. With **3-RSRCompServ** open, create a select query to select all the work order records for November 1, 2021 through November 7, 2021 by completing the following steps:
   a. Create a new query in Design view. Add the WorkOrders table to the query window and then close the Show Table dialog box. Drag down the bottom border of the table field list box until all the field names can be seen. If necessary, resize the query design grid.
   b. Double-click the WorkOrders table field list box title bar. This selects all the fields within the table.
   c. Position the mouse pointer within the selected field names in the table field list box and then drag the pointer to the first column in the query design grid. All the fields in the table are added to the query design grid.
   d. Click in the *Criteria* row in the *ServDate* column, type Between November 1, 2021 and November 7, 2021, and then press the Enter key.

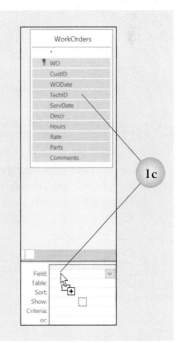

e. Run the query. The query results datasheet displays five records.

Work Order	Customer ID	WO Date	Technician ID	Service Date	Description	Hours	Rate	Parts
65012	1000	Mon, Nov 01 2021	11	Mon, Nov 01 2021	Biannual computer maintenance	1.25	50.00	$10.15
65013	1000	Tue, Nov 02 2021	10	Wed, Nov 03 2021	Replace keyboard	0.50	45.50	$42.75
65014	1005	Wed, Nov 03 2021	02	Wed, Nov 03 2021	Replace power supply	1.75	50.00	$62.77
65015	1008	Thu, Nov 04 2021	04	Fri, Nov 05 2021	Restore operating system	2.25	50.00	$0.00
65016	1010	Fri, Nov 05 2021	06	Fri, Nov 05 2021	Install upgraded video card	1.00	50.00	$48.75

1e

    f. It has been decided not to archive the *Comments* field data. Switch to Design view and then delete the *Comments* column from the query design grid.

2. Make a new table from the selected records and store the table in a history database to be used for archiving purposes by completing the following steps:

    a. If necessary, switch to Design view.

    b. Click the Make Table button in the Query Type group on the Query Tools Design tab.

    c. With the insertion point positioned in the *Table Name* text box, type Nov2021WO.

    d. Click the *Another Database* option and then click the Browse button.

    e. At the Make Table dialog box, navigate to the AL2C3 folder on your storage medium and then double-click the file named **3-RSRCompServHistory**.

    f. Click OK.

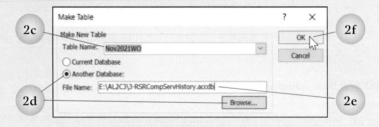

    g. Run the query.

    h. Click Yes at the Microsoft Access message box indicating that five rows are about to be pasted to a new table. You will verify that the query has run and created a table in later steps.

3. Save the query, typing Nov2021MakeTable as the query name, and then click OK.

4. Close the query.

5. Close **3-RSRCompServ**.

6. Open **3-RSRCompServHistory** and enable the content. Click OK to continue if a message appears stating that Access has to update object dependencies.

7. Open the Nov2021WO table in Datasheet view.

8. Review the records that were copied to the new table from the make table query that was run in Step 2g.

9. Close the table.

10. Close **3-RSRCompServHistory**.

Check Your Work

Creating a Delete Query

 Delete

A delete query is used to delete in one step a group of records that meet specified criteria. Use this action query when the records to be deleted can be selected using a criteria statement. Using a query to remove the records is more efficient and reduces the chances of removing a record in error, which can happen when records are deleted manually.

Quick Steps

Create Delete Query
1. Create query in Design view.
2. Add table to query.
3. Add fields to query design grid.
4. Enter criteria to select records.
5. Run query.
6. Switch to Design view.
7. Click Delete button.
8. Run query.
9. Click Yes.
10. Save query.

The make table query used in Activity 5a created a duplicate copy of the records in the new table. The original records still exist in the WorkOrders table. The make table query used to archive the records can be changed to a delete query and then used to remove the records from the original table.

In the Navigation pane, an action query name displays with a black exclamation mark next to an icon. The icon indicates the type of action that will be performed when the query is run.

Activity 5b **Deleting Records Using a Query**   Part 2 of 4

1. Open **3-RSRCompServ** and enable the content if necessary.
2. Right-click *Nov2021MakeTable* in the Queries group in the Navigation pane and then click *Design View* at the shortcut menu. Do not double-click to open/run it or Access will try to make a new table.
3. Click the Delete button in the Query Type group on the Query Tools Design tab to change the query to a Delete query.
4. Save the query with a new name, typing Nov2021Delete as the query name, and then click OK.
5. With the Query Tools Design tab selected, run the query.
6. At the Microsoft Access message box indicating that five rows are about to be deleted from the table and that the action cannot be reversed, click Yes to delete the selected records.

7. Close the query.
8. Open the WorkOrders table. Notice that no records have a service date before November 8, 2021.
9. Close the table.

Check Your Work

Creating an Append
Query

 Append

Quick Steps

Create Append Query
1. Create query in
   Design view.
2. Add table to query.
3. Add fields to query
   design grid.
4. Enter criteria to
   select records.
5. Run query.
6. Switch to Design
   view.
7. Click Append
   button.
8. Type table name.
9. Select destination
   database.
10. Click OK.
11. Run query.
12. Click Yes.
13. Save query.

Creating an Append Query

An append query is used to copy a group of records from one or more tables to the end of an existing table. Consider using an append query when a duplicate copy of records needs to be made. For example, in Activity 5a, the make table query was used to create a new table to store archived records. Once a table exists, use append queries to copy subsequent archived records to the end of the existing history table.

Click the Append button in the Query Type group on the Query Tools Design tab to open the Append dialog box. This dialog box opens with the same options as the Make Table dialog box, as shown in Figure 3.12. Click the *Table Name* option box arrow to choose from a list of existing tables in the specified database. The receiving table should have the same structure as the query from which the records are selected.

Figure 3.12 Append Dialog Box

Activity 5c Adding Records to a Table Using a Query

Part 3 of 4

1. With **3-RSRCompServ** open, open the Nov2021MakeTable query in Design view.
2. Modify the criteria to select work order records for the second week of November 2021 by completing the following steps:
   a. Expand the width of the *ServDate* column until the entire criteria statement can be seen.
   b. Click in the *Criteria* row in the *ServDate* field, insert and delete text as necessary to modify the criteria statement to read *Between #11/8/2021# And #11/14/2021#*, and then press the Enter key.

3. Click the Append button in the Query Type group on the Query Tools Design tab.
4. Since the query is being changed from a make table query, Access inserts the same table name and database that were used to create the table in Activity 5a. Click OK to accept the table name *Nov2021WO* and the file name *3-RSRCompServHistory*.

5. Save the query with a new name, typing Nov2021Append as the file name, and then click OK.
6. With the Query Tools Design tab selected, run the query.
7. Click Yes at the Microsoft Access message box indicating that three rows are about to be appended and that the action cannot be undone.
8. Close the query.
9. Close **3-RSRCompServ**.
10. Open **3-RSRCompServHistory**.
11. Open the Nov2021WO table and then print the datasheet in landscape orientation.
12. Close the table.
13. Close **3-RSRCompServHistory**.

Creating an Update Query

  Update

Quick Steps

Create Update Query
1. Create query in Design view.
2. Add table to query.
3. Add fields to query design grid.
4. Enter criteria to select records.
5. Run query.
6. Switch to Design view.
7. Click Update button.
8. Click in *Update To* row in field to be changed.
9. Type update expression.
10. Run query.
11. Click Yes.
12. Save query.

Creating an Update Query

When the change needed to a group of records can be selected in a query and is the same for all the records, instruct Access to modify the data using an update query. Making a global change using an update query is efficient and reduces the potential for errors that can occur from manually editing multiple records. Increase or decrease fields such as quotas, rates, and selling prices by adding or subtracting specific amounts or by multiplying by desired percentages.

Clicking the Update button in the Query Type group on the Query Tools Design tab causes an *Update To* row to appear in the query design grid. Click in the *Update To* row in the column to be modified and then type the expression that will change the field values as needed. Run the query to make the global change.

Exercise caution when running an action query because the queries make changes to database tables. For example, in Activity 5d, if the update query is run a second time, the rates for the Plan A service plans will increase another 6%. Once the rates have changed, they cannot be undone. Reversing the update would require creating and running, a mathematical expression in a new update query to remove 6% from the prices.

Activity 5d Changing Service Plan Rates Using an Update Query

Part 4 of 4

1. Open **3-RSRCompServ** and enable the content.
2. Open the FeesSCPlans table and review the current values in the *Rate* column. For example, notice that the current rate for Plan A's six-month term for one computer is $58.00. Close the table after reviewing the current rates.
3. Create an update query to increase the Plan A service contract rates by 4% by completing the following steps:
   a. Create a new query in Design view.
   b. Add the FeesSCPlans table to the query and then close the Show Table dialog box.
   c. Add the *Plan* field and *Rate* field to the query design grid.

d. Click in the *Criteria* row in the *Plan* column, type A, and then click in the *Criteria* row in the next column in the query design grid. ***Note: The AutoComplete feature will show a list of functions as soon as A is typed. Ignore the AutoComplete drop-down list, since a mathematical expression is not being entered for the criteria. However, pressing the Enter key causes Access to add*** Abs ***to the*** Criteria ***row. Clicking in another box in the query design grid will remove the AutoComplete drop-down list.***

e. Run the query. Review the four records shown in the query results datasheet.

f. Switch to Design view.
g. Click the Update button in the Query Type group on the Query Tools Design tab. Access adds a row labeled *Update To* in the query design grid between the *Table* row and *Criteria* row.
h. Click in the *Update To* row in the *Rate* column, type [Rate]*1.04, and then press the Enter key.

4. Save the query, typing RateUpdate as the query name, and then click OK.
5. Run the query. Click Yes at the Microsoft Access message stating that four rows are about to be updated.
6. Close the query.
7. Open the FeesSCPlans table. Notice that the rate values for the Plan A records have increased. For example, the value in the *Rate* column for Plan A's six-month term for one computer is now $60.32. Print the datasheet.
8. Close the table and then close **3-RSRCompServ**.

Chapter Summary

- Select queries extract records that meet specified criteria from a single table or multiple tables.

- A filter can be saved as a query by displaying the filter criteria in a Filter By Form window, clicking the Advanced button, and then clicking *Save As Query* at the drop-down list.

- Parameter queries prompt the user for the criteria by which to select records when the query is run. To create a parameter query, type a prompt message enclosed in square brackets in the *Criteria* row of the field to be used to select records.

- Changing the join property can alter the number of records that are displayed in the query results datasheet.

- An inner join displays records only if matching values are found in the joined field in both tables.

- A left outer join displays all the records from the left table and matching records from the related table. Empty fields display if no matching records exist in the related table.

- A right outer join displays all the records from the right table and matching records from the primary table. Empty fields display if no matching records exist in the primary table.

- Click the Show Table button in the Query Setup group on the Query Tools Design tab to add a table or query to the query window. Remove a table from a query by clicking a field within the table field list box and then pressing the Delete key or by right-clicking the table and clicking *Remove Table* at the shortcut menu.

- A self-join query is created by adding two copies of the same table to the query window and joining them by dragging the field with matching values from one table field list to the other.

- An alias is another name used to reference a table in a query. To assign an alias to a table, right-click the table name in the query window, click *Properties* at the shortcut menu, and then enter the alias for the table in the *Alias* property box in the Property Sheet task pane.

- A query that contains two tables that are not joined creates a cross product or Cartesian product query. This means that Access creates records for every possible combination from both tables—the results of which are generally not meaningful.

- A subquery is a query nested inside another query. Use subqueries to break down a complex query into manageable units. For example, a query with multiple calculations can be created by combining subqueries in which individual calculations are built individually. Another reason for using subqueries is to be able to reuse a smaller query in many other queries, avoiding the need to keep recreating the same structure.

- Using conditional logic requires Access to perform a calculation based on the outcome of a logical or conditional test.

- Create select queries on multiple-value fields using the same methods used for single-field criteria.

- Adding *.Value* at the end of a multiple-value field name in the *Field* row in the query design grid causes Access to place each field value in a separate row in the query results datasheet.

- A make table query creates a new table from selected records in the same database or another database using the structure defined in the query design grid.
- Delete a group of records in one step by creating and running a delete query. Add a group of records to the end of an existing table in the active database or another database by using an append query.
- An update query allows making a global change to records by entering an expression such as a mathematical formula in the query design grid.

Commands Review

FEATURE	RIBBON TAB, GROUP	BUTTON
advanced filter options	Home, Sort & Filter	
append query	Query Tools Design, Query Type	
create query in Design view	Create, Queries	
delete query	Query Tools Design, Query Type	
make table query	Query Tools Design, Query Type	
run query	Query Tools Design, Results	
show table	Query Tools Design, Query Setup	
update query	Query Tools Design, Query Type	

Performance Objectives

Upon successful completion of Chapter 4, you will be able to:

1. Create a custom form using Design view
2. Add fields to a form individually and as a group
3. Move, size, and format control objects
4. Change the tab order of fields
5. Add a tab control to a form and insert a subform
6. Add and format a calculation to a custom form
7. Adjust the alignment, sizing, and spacing of control objects
8. Add graphics to a form
9. Anchor a control object to a position in a form
10. Create a datasheet form and restrict form actions
11. Create a blank form
12. Add a list box and a combo box to a form
13. Locate a record using a wildcard character

A form provides an interface for data entry and maintenance that allows users to work more efficiently with data stored in the underlying tables. A form can also include fields from multiple tables, which allows data to be entered in one object and then used to update several tables. Generally, database designers provide forms for users to perform data maintenance and restrict access to tables, protecting the structure and integrity of the database. In this chapter, you will learn how to build custom forms.

 Data Files

Before beginning chapter work, copy the AL2C4 folder to your storage medium and then make AL2C4 the active folder.

The online course includes additional training and assessment resources.

Activity 1 Design and Create a Custom Form   7 Parts

> You will create a custom form using Design view that includes subforms in tabbed pages to provide a single object in which data stored in four tables can be entered, viewed, and printed.

Tutorial

Creating Forms
Using Design View

Form Design

Quick Steps

Start New Form Using Design View
1. Click Create tab.
2. Click Form Design button.

Creating Forms Using Design View

Access provides several tools that allow the user to build a form quickly, such as the Form tool, Split Form tool, and Form Wizard. A form generated using one of these tools can be modified in Layout view or Design view to customize the content, format, or layout. If several custom options are required in a form, begin in Design view and build the form from scratch. Click the Create tab and then click the Form Design button in the Forms group to begin a new form using the Design view window, shown in Figure 4.1.

In Design view, the form displays the *Detail* section, which is used to display fields from the table associated with the form. Objects are added to the form using buttons in the Controls, Header/Footer, and Tools groups on the Form Design Tools Design tab, shown in Figure 4.2.

Figure 4.1 Form Design View

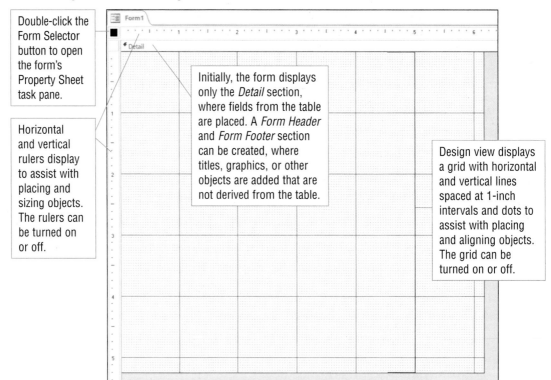

Double-click the Form Selector button to open the form's Property Sheet task pane.

Horizontal and vertical rulers display to assist with placing and sizing objects. The rulers can be turned on or off.

Initially, the form displays only the *Detail* section, where fields from the table are placed. A *Form Header* and *Form Footer* section can be created, where titles, graphics, or other objects are added that are not derived from the table.

Design view displays a grid with horizontal and vertical lines spaced at 1-inch intervals and dots to assist with placing and aligning objects. The grid can be turned on or off.

Figure 4.2 Controls, Header/Footer, and Tools Groups on the Form Design Tools Design Tab

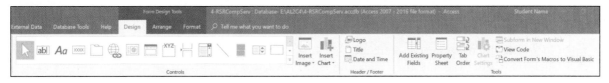

Understanding Bound, Unbound, and Calculated Control Objects

Three types of control objects can be created in a form. A control object in a form may be bound, unbound, or calculated. A bound control object draws and displays data in the control object from the field in the table associated with the control object. In other words, the content that is displayed in the control object in Form view is drawn from a field in a record in a table. An unbound control object is used to display text or graphics and does not rely on a table for its content. For example, a control object that contains an image to enhance the visual appearance of a form or a control object that contains the hours of business for informational purposes are both unbound control objects. A calculated control object displays the result of a mathematical formula. Totals and percentages are examples of calculated control objects.

Creating Titles and Label Control Objects

 Title

 Label

Quick Steps

Add Form Title
1. Open form in Design view.
2. Click Title button.
3. Type title text.
4. Press Enter key.

Add Label Control Object
1. Open form in Design view.
2. Click Label button.
3. Drag to create control object of specified height and width.
4. Type label text.
5. Press Enter key.

Click the Title button in the Header/Footer group on the Form Design Tools Design tab to display the *Form Header* and *Form Footer* sections and automatically insert a label control object with the name of the form inside the *Form Header* section. Select the text inside the title control object to type new text, delete existing text, or otherwise modify the default title text. Click the Label button in the Controls group to draw a label control object within any section in the form and then type descriptive or explanatory text inside the control object.

After creating a title or label control object, format the text using buttons in the Font group on the Form Design Tools Format tab. Move and resize the control object to reposition it on the form as needed.

Use the *Form Header* section to create control objects that are to be displayed at the top of the form when scrolling through records in Form view. This section is printed at the top of the page when a record or group of records is printed from Form view. Titles and company logos are generally placed in the *Form Header* section. Use the *Form Footer* section to create control objects that are to be displayed at the bottom of the form when scrolling through records in Form view. This section is printed at the end of a printout when a record or group of records is printed from Form view. Consider adding a creation date and/or revision number in the *Form Footer* section.

1. Open **4-RSRCompServ** and enable the content.
2. Click the Create tab and then click the Form Design button in the Forms group.

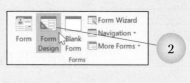

3. Add a title in the *Form Header* section of the form and then center the text within the control object by completing the following steps:

   a. With the Form Design Tools Design tab active, click the Title button in the Header/Footer group. Access displays the *Form Header* section above the *Detail* section and inserts a title control object with the text *Form1* selected.

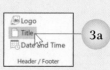

   b. With the insertion point positioned in the title control object and *Form1* selected, type Customer Data Maintenance Form. Notice that the background behind the title control object and the text of the title are both blue. The colors and fonts that appear in control objects are dependent on the current theme. The Office theme is the default for a database.

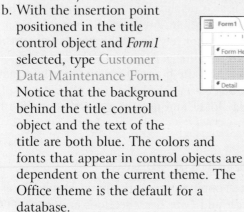

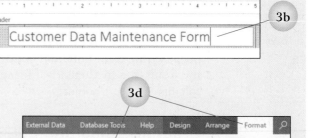

   c. Press the Enter key.
   d. With the title control object selected (as indicated by the orange border around the title text), click the Form Design Tools Format tab and then click the Center button in the Font group.
4. Scroll down the form until the *Form Footer* section is visible.
5. Position the mouse pointer at the bottom border of the form's grid until the pointer changes to an up-and-down-pointing arrow with a horizontal line in the middle and then drag down the bottom of the form to the 0.5-inch position on the vertical ruler.

6. Add one label control object at the left edge of the form footer that contains a revision number and another at the right edge of the form footer that contains your name by completing the following steps:

   a. Click the Form Design Tools Design tab and then click the Label button in the Controls group (third button from the left).
   b. Position the crosshairs with the label icon attached at the left side of the *Form Footer* section and then drag to draw a label control object of the approximate height and width shown at the right. When the mouse is released, the insertion point appears inside the label control object.

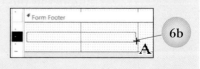

   c. Type Revision number 1.0 and then press the Enter key.

d. Create another label control object at the right side of the *Form Footer* section that is similar in height and width to the one shown below. Type your first and last names inside the label control object and then press the Enter key. Refer to Steps 6a–6c for help with this step.

e. Click in a blank area of the form to deselect the label control object.

7. Click the Save button on the Quick Access Toolbar, type CustMaintenance in the *Form Name* text box at the Save As dialog box, and then click OK.

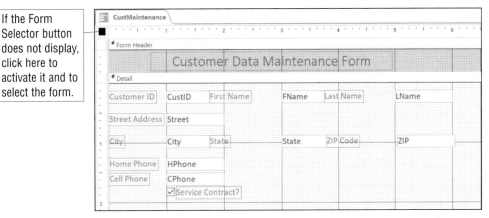

6d

7

> **Check Your Work**

 Tutorial

Adding Fields to a Form

 Form Selector

 Add Existing Fields

Quick Steps

Connect Table to Form
1. Open form in Design view.
2. Double-click Form Selector button.
3. Click Data tab in Property Sheet task pane.
4. Click arrow in *Record Source* property box.
5. Click table.
6. Close Property Sheet task pane.

Add Fields to Form
1. Click Add Existing Fields button.
2. Drag field name from Field List task pane to *Detail* section.
OR
1. Click first field name in Field List task pane.
2. Press and hold down Shift key.
3. Click last field name in Field List task pane.
4. Release Shift key.
5. Drag selected fields from Field List task pane to *Detail* section.

Adding Fields to a Form

Before fields can be added to the *Detail* section, a table must be connected to the form. Specify the table that will be connected with the form using the *Record Source* property box on the Data tab of the form's Property Sheet task pane. To do this, press the F4 function key or double-click the Form Selector button above the vertical ruler and left of the horizontal ruler; this opens the Property Sheet task pane at the right side of the work area. Click the Data tab, click the option box arrow in the *Record Source* property box, click the table at the drop-down list, and then close the Property Sheet task pane.

To add fields to the form, click the Add Existing Fields button in the Tools group on the Form Design Tools Design tab to open the Field List task pane. Select and drag individual or groups of field names from the Field List task pane to the *Detail* section of the form. Release the mouse button when the mouse pointer is near the location in the *Detail* section where the data will display.

For each field added to the form, Access inserts two control objects. A label control object is placed at the left of where the mouse button is released and a text box control object is placed at the right. The label control object contains the field name or a caption, if a caption was entered in the *Caption* property box. The text box control object is bound to the field and displays table data from the record in Form view. Figure 4.3 displays the fields from the Customers table that will be added to the CustMaintenance form in Activity 1b.

Figure 4.3 Fields from the Customers Table Added to the CustMaintenance Form in Activity 1b

1. With **4-RSRCompServ** open and the CustMaintenance form open in Design view, scroll to the top of the form in the work area.
2. Connect the Customers table to the form by completing the following steps:
   a. Double-click the Form Selector button (which displays as a black square) at the top of the vertical ruler and left of the horizontal ruler to open the form's Property Sheet task pane. *Note: If the Form Selector button does not display, click at the top of the vertical ruler and left of the horizontal ruler to display it.*
   b. Click the Data tab if necessary in the Property Sheet task pane.
   c. Click the option box arrow in the *Record Source* property box and then click *Customers* at the drop-down list.
   d. Close the Property Sheet task pane.

3. Add fields individually from the Customers table to the *Detail* section of the form by completing the following steps:
   a. Click the Add Existing Fields button in the Tools group on the Form Design Tools Design tab. The Field List task pane opens at the right side of the work area.
   b. Position the mouse pointer on the right border of the form's grid until the pointer changes to a left-and-right-pointing arrow with a vertical line in the middle and then drag the right edge of the form to the 6.5-inch position on the horizontal ruler.

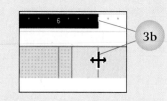

   c. If necessary, click *CustID* in the Field List task pane to select the field and then drag the field name to the *Detail* section. Release the mouse button when the pointer is near the top of the section at the 1-inch position on the horizontal ruler.

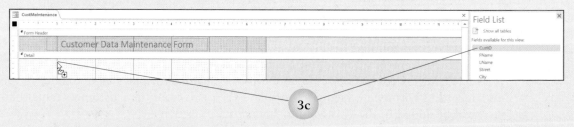

   d. Click to select *FName* in the Field List task pane and then drag the field to the same vertical position as *CustID* in the *Detail* section. Release the mouse button when the pointer is at the 3-inch position on the horizontal ruler. Some of the control objects may overlap. They will be adjusted in the following steps.
   e. Click to select *LName* in the Field List task pane and then drag the field to the same vertical position as *CustID* in the *Detail* section. Release the mouse button when the pointer is at the 5-inch position on the horizontal ruler.

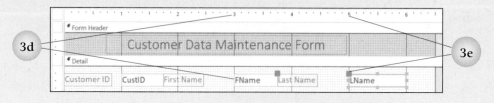

f. Drag the *Street* field from the Field List task pane to the *Detail* section below *CustID*. Release the mouse button when the pointer is at the 1-inch position on the horizontal ruler and approximately three rows of grid dots below *CustID*.

g. Drag the *City* field from the Field List task pane to the *Detail* section below *Street*. Release the mouse button when the pointer is at the 1-inch position on the horizontal ruler and approximately three rows of grid dots below *Street*.

h. Drag the *State* field from the Field List task pane to the *Detail* section at the same horizontal position as *City*. Release the mouse button when the pointer is at the 3-inch position on the horizontal ruler.

i. Drag the *ZIP* field from the Field List task pane to the *Detail* section at the same horizontal position as *City*. Release the mouse button when the pointer is at the 5-inch position on the horizontal ruler.

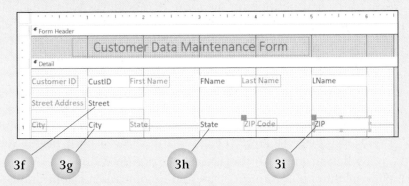

4. Add a group of fields from the Customers table to the *Detail* section of the form by completing the following steps:

a. Click the *HPhone* field name in the Field List task pane.

b. Press and hold down the Shift key, click the *ServCont* field name in the Field List task pane, and then release the Shift key. When the Shift key is held down, Access selects all the fields from the first field name clicked to the last field name clicked.

c. Position the mouse pointer within the selected group of fields in the Field List task pane and then drag the group to the *Detail* section below *City*. Release the mouse button when the pointer is at the 1-inch position on the horizontal ruler and approximately three rows of grid dots below *City*.

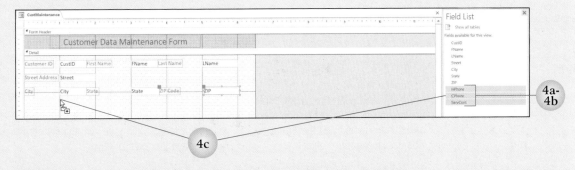

5. Click in any blank area of the form to deselect the group of fields.
6. Compare your form with the one shown in Figure 4.3 (on page 87).
7. Click the Save button on the Quick Access Toolbar.
8. Close the Field List task pane.

Tutorial

Moving, Resizing, and Formatting Control Objects

Quick Steps

Move Control Objects in Design View
1. Select control object.
2. Drag using orange border or move handle to new location.

Resize Control Objects in Design View
1. Select control object.
2. Drag middle top, bottom, left, or right sizing handle to resize height or width.
OR
1. Select control object.
2. Drag corner sizing handle to resize height and width at the same time.

Toggle Snap to Grid On or Off
1. Click Form Design Tools Arrange tab.
2. Click Size/Space button.
3. Click *Snap to Grid*.

 Size/Space

Hint Do not be overly concerned with exact placement and alignment when you begin to add fields to a form. Use the alignment and spacing tools in the Form Design Tools Arrange tab to assist with layout.

Moving and Resizing Control Objects

Once fields in a form have been placed, control objects can be moved or resized to change the layout. As shown in Activity 1b, Access places two control objects for each field in the form. A label control object, which contains the caption or field name, is placed left of a text box control object, which displays the field value from the record (or a blank entry box when a new record is added). Click the label control object or text box control object for the field to be moved or resized and the control object is surrounded by an orange border with eight handles. Access displays a large dark-gray square (called the *move handle*) at the top left of the label control object or text box control object for the selected field.

Point to the orange border around the selected control object until the mouse pointer displays with the four-headed arrow move icon attached and then drag the field to the new position on the form. Access moves both the label control and text box control objects to the new location. If a label control or text box control object is to be moved independently of its connected control object, point to the large dark-gray move handle at the top left of either control object and use it to drag the control object to the new position, as shown in Figure 4.4.

To resize a selected control object, point to one of the sizing handles (the small orange squares) on the border of the selected control object until the pointer displays with an up-and-down-pointing arrow, a left-and-right-pointing arrow, or a two-headed diagonal arrow. Drag the arrow to resize the height and/or width.

By default, the Snap to Grid feature is turned on in Design view. This feature pulls a control object to the nearest grid point when a control object is moved or resized. To move or resize a control object in small increments, turn off this feature. To do this, click the Form Design Tools Arrange tab, click the Size/Space button in the Sizing & Ordering group, and then click *Snap to Grid* at the drop-down list. Snap to Grid is a toggle feature, which means it is turned on or off by clicking the button.

Figure 4.4 Moving Control Objects

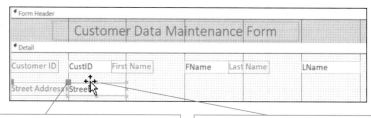

Point to the large dark-gray square (called the *move handle*) at the top left of the selected control object to move the selected *Street* text box control object independently of the *Street Address* label control object.

Point to the orange border and drag the object to the new location when the four-headed arrow move icon displays. The label control object containing the caption *Street Address* will move with the selected *Street* text box control object.

1. With **4-RSRCompServ** open and the CustMaintenance form open in Design view, preview the form to determine what control objects need to be moved or resized by clicking the View button in the Views group on the Form Design Tools Design tab. (Make sure to click the button and not the button arrow.)

2. The form is displayed in Form view and data from the first record is displayed in the text box control objects. Notice that some label control objects overlap text box control objects and that the street address in the first record is not entirely displayed.

3. Click the Design View button in the view area at the right side of the Status bar.

4. Move the control objects that overlap other control objects by completing the following steps:

   a. Click the *First Name* label control object.

   b. Point to the large dark-gray move handle at the top left of the selected label control object until the pointer displays with the four-headed arrow move icon attached and then drag right to the 2-inch position on the horizontal ruler. Notice that the connected *FName* text box control object does not move because the move handle is being used while dragging.

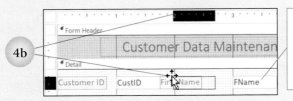

The connected *FName* text box control object does not move when the label control object is dragged using the move handle.

   c. Click the *Last Name* label control object and then use the move handle to drag right to the 4-inch position on the horizontal ruler.

   d. Move the *State* label control object right to the 2-inch position on the horizontal ruler.

   e. Move the *ZIP Code* label control object right to the 4-inch position on the horizontal ruler.

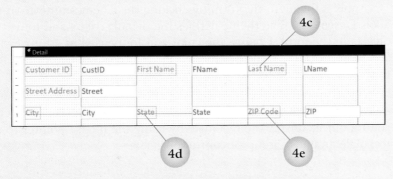

   f. Click in a blank area to deselect the *ZIP Code* label control.

5. Click the *Street* text box control object and then drag the right middle sizing handle right to the 3-inch position on the horizontal ruler.

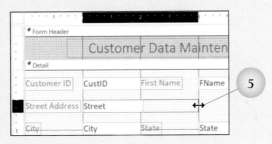

6. Click the *State* text box control object and then drag the right middle sizing handle left to the 3.5-inch position on the horizontal ruler. (Since the *State* field displays only two characters, this control object can be made smaller.)

7. Resize the *CustID* text box control object so the right edge of the object is at the 1.5-inch position on the horizontal ruler.

8. Click the *Service Contract?* label control object. Point to the selected control object's orange border (not on a sizing handle) and then drag the control object until the left edge is at the 3-inch position on the horizontal ruler, adjacent to the *HPhone* field. Notice that both the label control object and text box control object moved because the border was dragged rather than changed with the move handle.

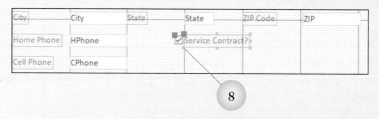

9. Deselect the *Service Contract?* control object.

10. Select the Cell Phone label control object and press the down arrow 3 times.

11. Save the form.

 Check Your Work

Formatting Control Objects

Quick Steps

Format Multiple Control Objects Using Shift Key

1. In Design view, click to select first control object.
2. Shift + click remaining control objects.
3. Click formatting options.
4. Deselect controls.

Format Multiple Control Objects Using Selection Rectangle

1. In Design view, position pointer above top left control to be formatted.
2. Drag down and right to draw rectangle around controls.
3. Release mouse.
4. Click formatting options.
5. Deselect controls.

 Themes

Use the buttons in the Font group on the Form Design Tools Format tab to change the font, font size, font color, background color, or alignment and to apply bold, italic, or underline formatting to the selected control object.

Apply a theme to the form by clicking the Themes button on the Form Design Tools Design tab. The theme defines the default colors and fonts for control objects in the form. Change the color and fonts used in a form with the Colors and Fonts buttons in the Themes group. A theme applied to one form is applied to all the forms and reports in the database.

Format multiple control objects at the same time by pressing and holding down the Shift key while clicking individual control objects. Select multiple control objects inside the rectangle by using the mouse pointer to draw a selection rectangle around a group of control objects.

1. With **4-RSRCompServ** open and the CustMaintenance form open in Design view, click the View button to preview the form in Form view.
2. Scroll through a few records in the form and then switch back to Design view.
3. Click the Design tab, click the Themes button in the Themes group and then click the Facet Theme.
4. Format multiple control objects using a selection rectangle by completing the following steps:
   a. Position the mouse pointer in the top left corner of the *Detail* section (above the *Customer ID* label control object), drag down and to the right until a rectangle has been drawn around all the control objects in the section, and then release the mouse button.

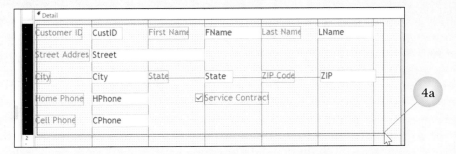

   b. Notice that all the control objects contained within the rectangle are selected.

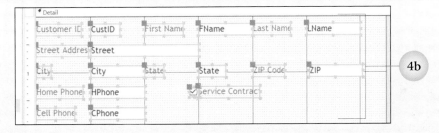

   c. Use the Font button in the Font group on the Form Design Tools Format tab to change the font to Candara.
   d. Use the Font Size button to change the font size to 10 points.
   e. Click in a blank area to deselect the control objects.
5. Apply formatting to multiple controls at once by completing the following steps:
   a. Click the *CustID* text box control object.
   b. Press and hold down the Shift key, click each of the other text box control objects in the *Detail* section, and then release the Shift key.

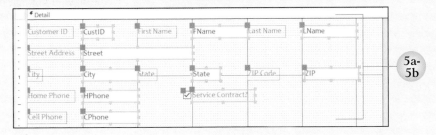

c. Click the Font Color button arrow in the Font group on the Form Design Tools Format tab and then click *Dark Red* at the color palette (first option in last row in the *Standard Colors* section).

d. Click the Bold button.

e. Click in a blank area to deselect the controls.

6. Click the Form Design Tools Design tab and then switch to Form view to view the formatting changes applied to the form. The zip code data remains red and does not change to dark red because of the formatting changes applied to the Format property in Chapter 1.

7. Switch to Design view and then save the form.

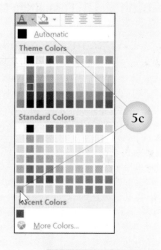

5c

Check Your Work

 Tutorial

Changing the Tab Order of Fields

 Tab Order

Changing the Tab Order of Fields

The term *tab order* refers to the order that fields are selected when the Tab key is pressed while entering data in Form view. Data does not need to be entered into a record in the order the fields are presented. Click the Tab Order button in the Tools group on the Form Design Tools Design tab to open the Tab Order dialog box, shown in Figure 4.5.

The order of the fields in the *Custom Order* list box in the Tab Order dialog box is the order in which the fields will be selected when the Tab key is pressed in a record in Form view. To change the tab order of the fields, position the pointer in the field selector bar next to the field name to be moved until the pointer displays as a black right-pointing arrow and then click to select the field. Drag the selected field up or down to the appropriate position. Click OK to accept the changes and close the Tab Order dialog box. To quickly set the tab order as left-to-right and top-to-bottom, click the Auto Order button at the bottom of the Tab Order dialog box.

Quick Steps

Change Tab Order of Fields

1. Open form in Design view.
2. Click Tab Order button.
3. Click in field selector bar next to field name.
4. Drag field to new location.
5. Repeat Steps 3–4 as required.
6. Click OK.

Figure 4.5 Tab Order Dialog Box

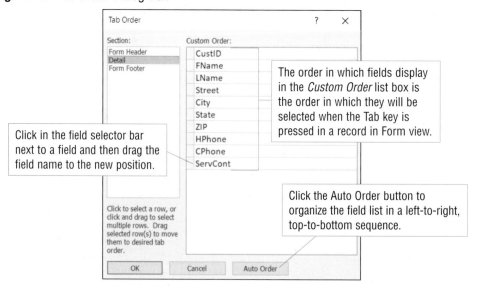

1. With **4-RSRCompServ** open and the CustMaintenance form open in Design view, click the View button to display the form in Form view.
2. With the insertion point positioned in the *CustID* field in the first record in the table, press the Tab key seven times. As the Tab key is pressed, notice that the fields are selected in a left-to-right, top-to-bottom sequence.
3. With the insertion point positioned in the *HPhone* field, press the Tab key. Notice that the selected field moves down to the *CPhone* field instead of right to the *ServCont* field.
4. With the insertion point in the *CPhone* field, press the Tab key to move to the *ServCont* field.
5. Switch to Design view.
6. Change the tab order of the fields so the *ServCont* field is selected after the *HPhone* field by completing the following steps:
   a. Click the Tab Order button in the Tools group on the Form Design Tools Design tab.
   b. At the Tab Order dialog box, hover the mouse pointer over the field selector bar next to *ServCont* until the pointer displays as a black right-pointing arrow.
   c. Click to select the field.
   d. With the pointer now displayed as a white arrow in the field selector bar, drag up *ServCont* until the black horizontal line indicating where the field will be moved is positioned between *HPhone* and *CPhone* in the *Custom Order* list and then release the mouse button.
   e. Click OK. Note that since the tabs follow the left-to-right, top-to-bottom sequence, this tab order could have been created by clicking the Auto Order button in the Tab Order dialog box.
7. Switch to Form view.
8. Press the Tab key nine times to move through the fields in the first record. Notice that when the *HPhone* field is reached and the Tab key is pressed, the *ServCont* field becomes active instead of the *CPhone* field.
9. Switch to Design view.
10. Save the form.

Adding a Tab Control Object to a Form

Adding a Tab Control Object to a Form

A tab control object is an object used to add pages to a form. Each page displays with a tab at the top. When viewing the form, click the page tab to display the contents of the page within the tab control object. Add a tab control object to a form to organize fields in a large table into smaller, related groups or to insert multiple subforms that display on separate pages within the tab control object.

Examine the tab control object that will be created in Activities 1f and 1g, as shown in Figure 4.6. The tab control object contains three pages. The tabs across the top display the captions assigned to the individual pages.

In Activities 1f and 1g, a subform will be created on each page within the tab control object to display fields from a related table. When completed, the CustMaintenance form will be used to enter or view customer-related data that includes fields from four tables.

**Add Tab Control
Object to Form**

1. Open form in Design view.
2. Click Tab Control button.
3. Position crosshairs in *Detail* section at specified location.
4. Drag down and right to draw object.
5. Release mouse.

Change Page Caption

1. Click tab in tab control object.
2. Click Property Sheet button.
3. Click in *Caption* property box.
4. Type text.
5. Close Property Sheet task pane.

**Add Page to
Tab Control**

1. Right-click existing tab in tab control object.
2. Click *Insert Page*.

**Delete Page from
Tab Control**

1. Right-click tab of page to delete.
2. Click *Delete Page*.

Figure 4.6 Tab Control Object with Three Pages Created in Activities 1f and 1g

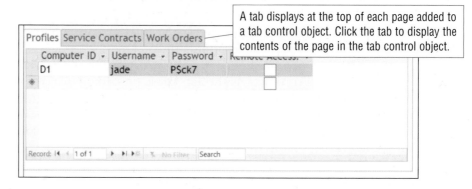

A tab displays at the top of each page added to a tab control object. Click the tab to display the contents of the page in the tab control object.

Figure 4.7 New Tab Control Object with Two Pages

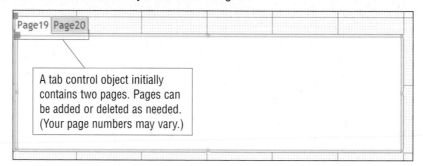

A tab control object initially contains two pages. Pages can be added or deleted as needed. (Your page numbers may vary.)

Tab Control

To add a tab control object to a form, click the Tab Control button in the Controls group on the Form Design Tools Design tab. Position the crosshairs with the tab control icon attached in the top left of the area of the *Detail* section where the tab control is to begin and then drag down and to the right to draw the control. When the mouse button is released, the tab control object initially displays with two pages, as shown in Figure 4.7.

Change the text displayed in the tab at the top of the page by editing the text in the *Caption* property box in the page's Property Sheet task pane. Add fields or create subforms on each page as needed. Add a page to the tab control object by right-clicking an existing tab in the tab control object and then clicking *Insert Page* at the shortcut menu. Remove a page from the tab control object by right-clicking the tab to be deleted and then clicking *Delete Page* at the shortcut menu.

Tutorial

Adding a Subform

Adding a Subform

Use the Subform/Subreport button in the Controls group on the Form Design Tools Design tab to add a subform to a form. Create a subform to display fields from another related table within the existing form. The form that a subform is created in is called the *main form*.

Adding a related table as a subform creates a control object within the main form that can be moved, formatted, and resized independently of other objects. The subform displays as a datasheet within the main form in Form view. Data can be entered or updated in the subform while the main form is being viewed. Before clicking the Subform/Subreport button, make sure the Use Control Wizards button in the Controls group is toggled on so the subform can be added using the Subform Wizard; the first Subform Wizard dialog box is shown in

Subform/
Subreport

Use Control
Wizards

Figure 4.8. The Use Control Wizards button displays with a gray background when the feature is active.

Quick Steps

Create Subform
1. Click page tab in tab control object.
2. Make sure Use Control Wizards is toggled on.
3. Click More button in Controls group.
4. Click Subform/ Subreport button.
5. Click crosshairs inside selected page.
6. Click Next.
7. Choose table and fields.
8. Click Next.
9. Click Next.
10. Click Finish.
11. Delete subform label control object.
12. Move and resize subform object as required.

A subform is stored as a separate object outside the main form. An additional form name (with *subform* at the end of it) will appear in the Navigation pane when the Subform Wizard is finished. Do not delete a subform object from the Navigation pane. If the subform object is deleted, the main form will no longer display the fields from the related table in the tab control object.

Figure 4.8 First Subform Wizard Dialog Box

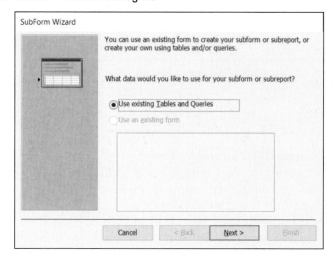

Activity 1f Adding a Tab Control Object and a Subform

Part 6 of 7

1. With **4-RSRCompServ** open and the CustMaintenance form open in Design view, add a tab control object to the form by completing the following steps:
   a. Click the Tab Control button in the Controls group on the Form Design Tools Design tab. The Tab Control button is the fifth button from the left in the Controls group.
   b. Position the crosshairs with the tab control icon attached at the left edge of the grid in the *Detail* section at the 2-inch position on the vertical ruler. Drag down to the 4-inch position on the vertical ruler and right to the 6-inch position on the horizontal ruler and then release the mouse button.

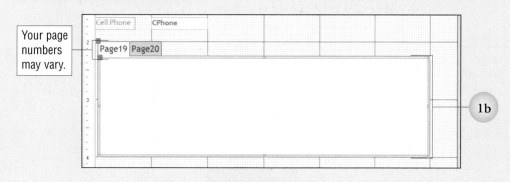

2. Change the page caption and add a subform to the first page within the tab control object by completing the following steps:

a. Click the first tab in the tab control object that displays *Pagexx*, where *xx* is the page number to select the page. (For example, click *Page19* in the image shown above.)

b. Click the Property Sheet button in the Tools group.

c. Click the Format tab in the Property Sheet task pane, click in the *Caption* property box, type Profiles, and then close the Property Sheet task pane. The tab displays the caption text in place of *Pagexx*.

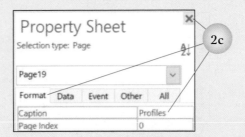

d. By default, the Use Control Wizards feature is toggled on in the Controls group. Click the More button in the Controls group to expand the Controls group and view two rows of buttons and the Controls drop-down list. View the current status of the Use Control Wizards button. If the button displays with a gray background, the feature is active. If the button is gray, click in a blank area to remove the expanded Controls list. If the feature is not active (displays with a white background), click the Use Control Wizards button to turn it on.

e. Click the More button in the Controls group and then click the Subform/Subreport button at the expanded Controls list.

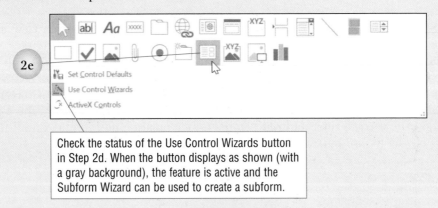

Check the status of the Use Control Wizards button in Step 2d. When the button displays as shown (with a gray background), the feature is active and the Subform Wizard can be used to create a subform.

f. Move the crosshairs with the subform icon attached to the Profiles page in the tab control object. The background of the page turns black. Click the mouse to start the SubForm Wizard.

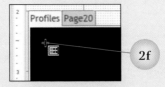

g. At the first SubForm Wizard dialog box with *Use existing Tables and Queries* already selected, click the Next button.

h. At the second SubForm Wizard dialog box, select the table and fields to be displayed in the subform by completing the following steps:

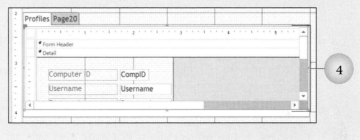

1) Click the *Tables/Queries* option box arrow and then click *Table: Profiles* at the drop-down list.
2) Move all the fields except *CustID* from the *Available Fields* list box to the *Selected Fields* list box.
3) Click the Next button.

i. At the third SubForm Wizard dialog box with *Show Profiles for each record in Customers using CustID* selected, click the Next button. Since a relationship has been created between two tables with *CustID* as the joined field, Access knows the field that links the records in the main form with the subform.

j. Click the Finish button at the last SubForm Wizard dialog box to accept the default subform name *Profiles subform*.

3. Access creates the subform within the active page in the tab control with a label control object above it. Click the label control object displaying the text *Profiles subform* to select the object and then press the Delete key.

4. Click the edge of the subform control object to display the orange border and sizing handles and then use the techniques learned in Activity 1c to move and resize the object so that the subform fills the tab control object as shown at the right.

5. Click in a blank area outside the grid to deselect the subform control object and then switch to Form view. Notice that the subform displays as a datasheet within the tab control object in the CustMaintenance form.

6. In the field names row in the datasheet, position the mouse pointer on each boundary line that separates the columns and then double-click to adjust each column's width to best fit.

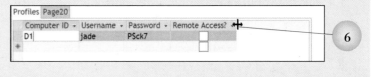

7. Notice that two sets of navigation buttons are displayed: one at the bottom of the main form (just above the Status bar) and another at the bottom of the datasheet in the subform. Use the navigation buttons at the bottom of the main form to scroll through a few records and watch the fields update in both the main form and subform upon moving to the next customer record.

8. Switch to Design view.

9. Save the form.

10. In the Navigation pane, notice that a form object exists with the name *Profiles subform*. Subforms are separate objects within the database. If the main form is closed, the subform can still be opened to edit data.

Check Your Work

In Design view, the control objects within the subform display one below another, but in Form view, the subform displays using a datasheet layout. Change the setting of the *Default View* property box in the subform's Property Sheet task pane to *Single Form* to match the layout of the controls in Design view to the layout of the fields in Form view. The fields display one below another in a single column in Form view.

To do this, open the subform's Property Sheet task pane by double-clicking the Form Selector button at the top of the vertical ruler and left of the horizontal ruler in the subform control object in Design view. Click in the *Default View* property box on the Format tab, click the option box arrow, and then click *Single Form* at the drop-down list.

Activity 1g Adding More Subforms and Adding a New Page to the Tab Control Object   Part 7 of 7

1. With **4-RSRCompServ** open and the CustMaintenance form open in Design view, change the caption for the second page in the tab control object to *Service Contracts* by completing steps similar to those in Steps 2a–c of Activity 1f.
2. With the Service Contracts page selected in the tab control object, add a subform to display the fields from the ServiceContracts table on the page by completing the following steps:
    a. Click the More button in the Controls group and then click the Subform/Subreport button in the expanded Controls group.
    b. Click inside the selected Service Contracts page in the tab control object.
    c. Click the Next button at the first SubForm Wizard dialog box.
    d. At the second SubForm Wizard dialog box, change the table displayed in the *Tables/Queries* option box to *Table: ServiceContracts*.
    e. Move all the fields from the table except the *CustID* field to the *Selected Fields* list box.
    f. Click the Next button.
    g. At the third SubForm Wizard dialog box with *Show ServiceContracts for each record in Customers using CustID* selected, click the Next button.
    h. Click the Finish button at the last SubForm Wizard dialog box to accept the default subform name *ServiceContracts subform*.
3. Select and then delete the label control object above the subform (which displays the text *ServiceContracts subform*).
4. Click the subform control object to display the orange border and sizing handles and then move and resize the form as shown below.

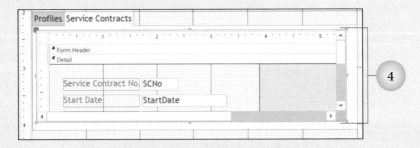

5. Deselect the subform control object and then switch to Form view.

6. Click the Service Contracts
   tab and then adjust the
   width of each column in
   the datasheet to best fit. If
   necessary, use the scroll bar to
   access the right border for the *PlanID* field.

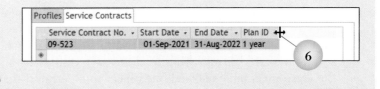

7. Switch to Design view and then save the form.
8. Add a new page to the tab control object and add a subform to display selected fields from
   the WorkOrders table in the new page by completing the following steps:
   a. Right-click the Service Contracts tab and then click *Insert Page* at the shortcut menu.
   b. With the new page already selected, display the Property Sheet task pane, type Work
      Orders as the page caption, and then close the Property Sheet task pane.
   c. Click the More button in the Controls group and then click the Subform/Subreport
      button. Click inside the selected Work Orders page. Create a subform to display selected
      fields from the WorkOrders table by completing the following steps:
      1) Click the Next button at the first SubForm Wizard dialog box.
      2) At the second SubForm Wizard dialog box, change the table displayed in the
         Tables/Queries option box to *Table: WorkOrders*.
      3) Move the following fields from the *Available Fields* list box to the *Selected Fields* list box:
         WO
         WODate
         Descr
      4) Click the Next button.
      5) Click the Next button at the third SubForm Wizard dialog box.
      6) Click the Finish button at the last SubForm Wizard dialog box to accept the default
         subform name.
9. Select and then delete the label control object above the subform (which displays the text
   WorkOrders subform).
10. Click the subform control object to display the orange border and sizing handles and then
    move and resize the form as shown below.
11. Access automatically extends the width of the form and widens the tab control object if a
    table with many fields is added to a subform. If necessary, select the tab control object
    and decrease the width so the right edge of the tab control is at the 6-inch position on the
    horizontal ruler. If necessary, decrease the width of the form so the right edge of the grid is
    at the 6.5-inch position on the horizontal ruler. ***Hint: If Access resizes the tab control object
    to the edge of the form, the grid may have to be temporarily widened to see the middle sizing
    handle at the right edge of the tab control object.***

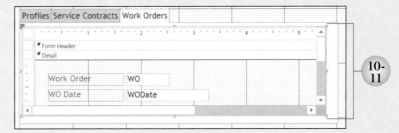

12. Deselect the subform control object and then switch to Form view.
13. While viewing the form, you decide the title would look better if it was not centered. Switch
    to Design view, click the Title control object in the *Form Header* section, click the Form
    Design Tools Format tab, and then click the Align Left button in the Font group.

14. Click the Form Design Tools Design tab and then switch to Form view.
15. Click the Work Orders tab and adjust the width of each column in the datasheet to best fit. Compare the CustMaintenance form with the one shown in Figure 4.9. (Notice that adding the tab control with a separate page displaying a subform for each table related to the Customers table allowed one object to be created that can be used to view and update fields in multiple tables.)
16. Print only the selected record. To do this, open the Print dialog box, click *Selected Record(s)* in the *Print Range* section, and then click OK.
17. Save and then close the CustMaintenance form.

Figure 4.9 Completed CustMaintenance Form

<table>
<tr><td>

Activity 2 **Create a New Form with Calculations and Graphics**   **5 Parts**

You will create a new form using the Form Wizard, add two calculations to the form, use features that assist with alignment and spacing of multiple control objects, and add graphics to the form.

</td></tr>
</table>

 Tutorial

Adding Calculations
to a Form

 Text Box

Quick Steps

**Add Calculated
Control Object to
Form**

1. Open form in Design view.
2. Click Text Box button.
3. Position crosshairs in *Detail* section at specified location.
4. Drag to create control object of required height and width.
5. Release mouse.
6. Click in text box control object.
7. Type formula.
8. Press Enter key.
9. Delete text in label control object.
10. Type label text and press Enter key.

Adding Calculations to a Form

To display a calculated value in a form, create a query that includes a calculated column and then create a new form based on that query. Alternatively, create a calculated control object in an existing form using Design view.

To do this, click the Text Box button in the Controls group on the Form Design Tools Design tab and then drag the crosshairs in the *Detail* section to create a control object the approximate height and width required to show the calculation. Access displays a text box control object with *Unbound* inside the object and a label control object to the left displaying *Textxx* (where *xx* is the text box control object number). A calculated control object is considered as unbound because the data displayed in the control is not drawn from a stored field value in a record. Click inside the text box control object (*Unbound* disappears) and then type the formula, beginning with an equals sign. For example, the formula *=[Hours]*[Rate]* multiplies the value in the *Hours* field times the value in the *Rate* field. The field names in a formula are enclosed in square brackets.

Edit the label control object next to the calculated control to add a descriptive name that identifies the calculated value. Open the Property Sheet task pane for the calculated control object and then apply the Fixed, Standard, or Currency format, as appropriate for the calculated value. Since the data displayed in a calculated control object is based on a formula, there is no need to tab to this control object when entering data or moving through a record. Change the *Tab Stop* property box on the Other tab of the Property Sheet task pane from *Yes* to *No* to avoid stopping at any text box control object.

Activity 2a   Adding and Formatting Calculated Control Objects   Part 1 of 5

1. With **4-RSRCompServ** open, use the Form Wizard to create a new form based on the WorkOrders table by completing the following steps:
   a. Click *WorkOrders* in the Tables group in the Navigation pane and then click the Create tab.
   b. Click the Form Wizard button in the Forms group.
   c. With *Table: WorkOrders* selected in the *Table/Queries* option box, complete the steps in the Form Wizard as follows:
      1) Move the *WO, Descr, ServDate, Hours, Rate,* and *Parts* fields from the *Available Fields* list box to the *Selected Fields* list box and then click the Next button.
      2) With *Columnar* selected as the layout option, click the Next button.
      3) With *WorkOrders* the default text in the *What title do you want for your form?* text box, click *Modify the form's design* at the last dialog box and then click the Finish button.
2. With the WorkOrders form displayed in Design view, add a calculated control object to display the total labor for the work order by completing the following steps:

a. Position the pointer on the top border of the gray *Form Footer* section bar until the pointer displays as an up-and-down-pointing arrow with a horizontal line in the middle and then drag down just below the 3-inch position on the vertical ruler. (The 3-inch mark will not appear until the border is dragged below it and the mouse button is released.) Doing this creates more grid space in the *Detail* section so that controls can be added.

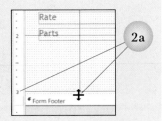

b. Click the Text Box button in the Controls group.

c. Position the crosshairs with the text box icon attached below the *Parts* text box control, drag to create an object of the approximate height and width shown at the right, and then release the mouse button.

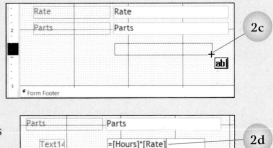

d. Click in the text box control (which displays *Unbound*) and then type =[Hours]*[Rate].

e. Press the Enter key.

f. With the calculated control object selected, click the Property Sheet button in the Tools group. With the Format tab in the Property Sheet task pane active, click the option box arrow in the *Format* property box, click *Standard* at the drop-down list, and then close the Property Sheet task pane. By default, calculated values display right-aligned in Form view.

g. Click to select the label control object to the left of the calculated control object (which displays *Textxx* [where *xx* is the text box label number]). Click in the selected label control object a second time to place the insertion point. Delete *Textxx*, type Total Labor, and then press the Enter key. Notice that the label control automatically expands to accommodate the width of the typed text.

3. With the Form Design Tools Design tab selected, click the View button to display the form in Form view and then scroll through a few records to view the calculated field. Do not be concerned with the size, position, alignment, and/or spacing of the controls; the format will be fixed in a later activity.

4. Switch to Design view and then save the form.

5. A calculated control object can be used as a field in another formula. To do this, reference the calculated object in the formula by its Name property enclosed in square brackets. Change the name for the calculated object created in Step 2 to a more descriptive name by completing the following steps:

a. Click the calculated control object (which displays the formula *=[Hours]*[Rate]*).

b. Click the Property Sheet button in the Tools group.

c. Click the Other tab in the Property Sheet task pane.

d. Select and delete the existing text (which displays *Textxx* [where *xx* is the text box number]) in the *Name* property box.

e. Type LaborCalc, press the Enter key, and then close the Property Sheet task pane.

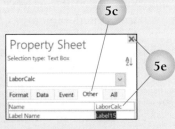

6. Add another calculated control object to the form to include labor and parts and determine the total value for the work order by completing the following steps:

a. Click the Text Box button in the Controls group.

b. Position the crosshairs with the text box icon attached below the calculated control created in Step 2, drag to create an object the approximate height and width as the first calculated control, and then release the mouse button.

c. Click in the text box control (which displays *Unbound*), type =[LaborCalc]+[Parts], and then press the Enter key.

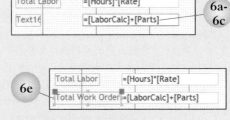

d. Apply the Currency format to the calculated control. Refer to Step 2g for assistance with this step.

e. Type Total Work Order in the label control object. Refer to Step 2h for assistance with this step.

7. Remove the tab stops from the two new calculated control objects by completing the following steps:

a. Press and hold the Shift key, use the left mouse button to click the calculated control objects that display the formulas *=[Hours]*[Rate]* and *=[LaborCalc]+[Parts]*, and then release the Shift key.

b. Click the Property Sheet button in the Tools group.

c. Click the Other tab in the Property Sheet task pane.

d. Double-click *Yes* in the *Tab Stop* property box to change it to *No*.

e. Close the Property Sheet task pane.

8. Save the form. Display the form in Form view and then scroll through a few records to view the calculations. Tab through a record and notice that when the Tab key is pressed after the *Parts* field, the insertion point does not stop at *Total Labor* or *Total Work Order* but instead moves to the next record.

9. Switch to Design view.

 Check Your Work

 Tutorial

Sizing, Aligning, and Spacing Multiple Control Objects

 Align

Adjusting Control Objects for Consistency in Appearance

When working in Design view, use the tools that Access provides for positioning, aligning, sizing, and spacing multiple controls to create forms with a consistent and professional appearance. Locate these tools using the Size/Space button and Align button in the Sizing & Ordering group on the Form Design Tools Arrange tab.

Aligning Multiple Control Objects

Quick Steps

Align Multiple Control Objects
1. In Design view, select control objects.
2. Click Form Design Tools Arrange tab.
3. Click Align button.
4. Click option at drop-down list.
5. Deselect control objects.

The options in the Align button drop-down list in the Sizing & Ordering group on the Form Design Tools Arrange tab are shown in Figure 4.10. Use these options to align multiple selected control objects at the same horizontal or vertical position and avoid having to adjust each control object individually.

Figure 4.10 Options in the Align Button Drop-Down List

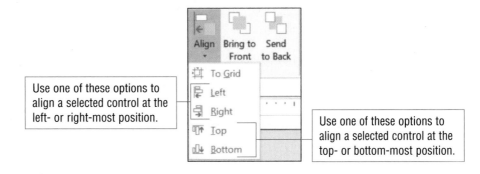

Adjusting the Sizing and Spacing between Control Objects

 Size/Space

Quick Steps

Adjust Spacing between Control Objects
1. In Design view, select control objects.
2. Click Form Design Tools Arrange tab.
3. Click Size/Space button.
4. Click option in *Spacing* section.
5. Deselect controls.

The options provided by the Size/Space button in the Sizing & Ordering group on the Form Design Tools Arrange tab are shown in Figure 4.11. Use these options to assist with the consistent sizing of and spacing between control objects. Use options in the *Size* section of the drop-down list to adjust the height or width to the tallest, shortest, widest, or narrowest of the selected control objects. Use options in the *Spacing* section to adjust the horizontal and vertical spacing between control objects, increase or decrease the space, or provide equal spaces between all the selected control objects.

Using these tools is helpful when creating a new form by adding control objects manually to the grid or after editing an existing form because the space between control objects can be changed easily after objects are added or deleted. To adjust the spacing by moving individual control objects would be too time consuming.

Figure 4.11 Size and Spacing Options in the Size/Space Button Drop-Down List

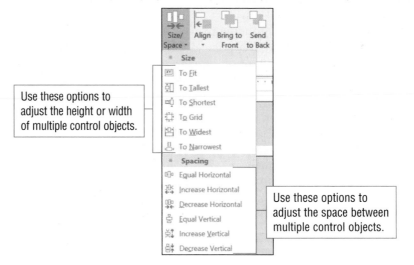

Use these options to adjust the height or width of multiple control objects.

Use these options to adjust the space between multiple control objects.

Activity 2b Sizing, Aligning, and Spacing Multiple Control Objects   Part 2 of 5

1. With **4-RSRCompServ** open and the WorkOrders form open in Design view, change the font of the text in the *Title* control object to 16 pts and then change it to read *Work Orders with Calculations*. Widen the *Title* control object to fit the title on one line.
2. With the title control object still selected, position the mouse pointer on the orange border until the pointer changes to the four-headed arrow move icon and then drag the control object to the approximate center of the *Form Header* section.

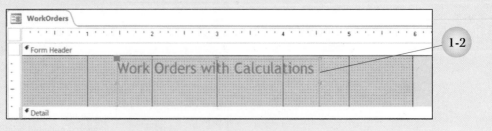

3. Point to the bottom gray border in the *Form Footer* section bar until the pointer displays as an up-and-down-pointing arrow with a horizontal line in the middle and then drag down approximately 0.5 inch to create space in the *Form Footer* section. Create a label control object with your name in the center of the *Form Footer* section.

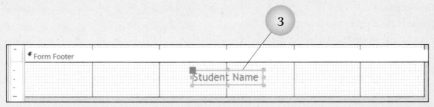

4. Click to select the *Descr* text box control object and then drag the right middle sizing handle left until the control object is resized to approximately the 4.5-inch position on the horizontal ruler.
5. Press and hold down the Shift key, use the left mouse button to click the six text box control objects for the fields above the two calculated control objects, and then release the Shift key. Click the Form Design Tools Arrange tab, click the Size/Space button in the Sizing & Ordering group, and then click *To Widest* at the drop-down list. The six text box control objects are now all the same width. (The width is set to fit the widest selected control object).

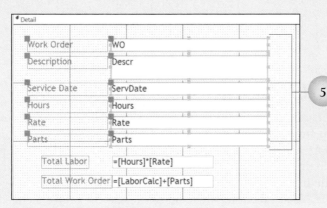

6. Click in a blank area to deselect the control objects.
7. Use the Align button to align multiple control objects by completing the following steps:
   a. Draw a selection rectangle around all the label control objects at the left side of the form. This selects all eight label control objects.
   b. Click the Align button in the Sizing & Ordering group and then click *Left* at the drop-down list. All the label control objects align at the left edge of the left-most control object.
   c. Deselect the control objects.
   d. Draw a selection rectangle around all the text box control objects at the right of the form to select all eight text box control objects.
   e. Click the Align button in the Sizing & Ordering group and then click *Right* at the drop-down list. All the control objects align at the right edge of the right-most control object.
   f. Deselect the control objects.

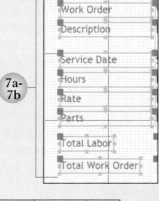

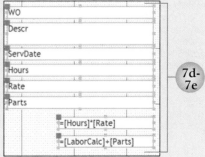

8. Adjust the vertical spaces between control objects to make all the control objects in the *Detail* section equally spaced by completing the following steps:

   a. Draw a selection rectangle around all the control objects in the *Detail* section.

   b. Click the Size/Space button and then click *Equal Vertical* in the *Spacing* section of the drop-down list. All the control objects are now separated by the same amount of vertical space.

   c. Deselect the control objects.

9. Save the form.

10. Display the form in Form view and then scroll through a few records to view the revised alignment and spacing. The numeric fields will be formatted in Activity 2e to align the numbers correctly.

11. Switch to Design view.

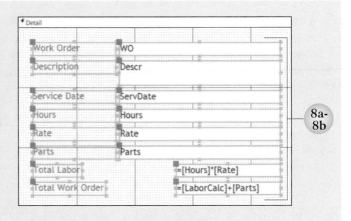

8a-8b

Check Your Work

Adding Graphics to a Form

Tutorial

Adding and Anchoring Graphics to a Form

 Logo

 Insert Image

 Line

Quick Steps

Add Image to Form
1. Open form in Design view.
2. Click Insert Image button.
3. Click *Browse* and navigate to folder.
4. Double-click image.
5. Click and drag crosshairs.
6. Move and resize.
7. If necessary, display Property Sheet task pane and change Size Mode property.

A picture that is saved in a graphic file format can be added to a form using the Logo button or Insert Image button in the Controls group on the Form Design Tools Design tab. Click the Logo button and the Insert Picture dialog box opens. Navigate to the drive and/or folder in which the graphic file is stored and then double-click the image file name. Access automatically adds the image to the left side of the *Form Header* section. Move and/or resize the image as needed. Access supports the BMP, GIF, JPEG, JPG, and PNG graphic file formats for a logo control object.

Use the Insert Image button to place the picture in another section or draw a larger control object to hold the picture. Click the Insert Image button and then click *Browse* at the drop-down list to open the Insert Picture dialog box. Navigate to the drive and/or folder in which the graphic file is stored and then double-click the image file name. Position the crosshairs with the image icon attached at the location in the form where the image is to be placed and then drag the crosshairs to draw a control object of the approximate height and width desired. Access supports the GIF, JPEG, JPG, and PNG file formats for an image control object.

Use the Line button in the Controls group to draw horizontal or vertical lines in a form. Press and hold down the Shift key while dragging to draw a straight line. Once the line has been drawn, use the Shape Outline button in the Control Formatting group on the Form Design Tools Format tab to modify the line thickness, type, and color.

Online images can be added to a form. Access does not provide an online pictures button in the Controls group. However, an online picture can be inserted in a Microsoft Word document and then saved to a folder using standard Windows commands. Use the Insert Image button to insert the saved image.

1. With **4-RSRCompServ** open and the WorkOrders form open in Design view add an image to the WorkOrders form by completing the following steps:
   a. Click the Insert Image button in the Controls group, click the *Browse* option, navigate to your AL2C4 folder, and then double click *CompRepair*.

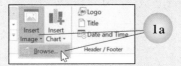

   b. Position the crosshairs with the picture icon attached to the right of the text box control objects, drag to create an object of the approximate height and width shown, and then release the mouse button. Access inserts the image and displays the orange border with selection handles.

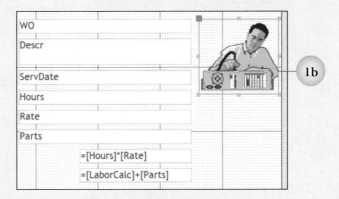

2. Click the Property Sheet button in the Tools group. If necessary, make Format the active tab in the Property Sheet task pane. Click in the *Size Mode* property box (displays *Zoom*), click the option box arrow that appears, and then click *Stretch* at the drop-down list. Close the Property Sheet task pane. Changing the *Size Mode* property box to display *Stretch* instructs Access to resize the image to fit the height and width of the control object. Note that using *Stretch* may skew the appearance of the image.

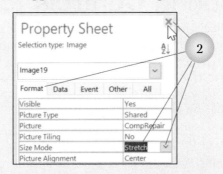

3. Deselect the control object containing the image and then display the form in Form view. Add a line below the title to improve the appearance of the form. Draw and modify the line by completing the following steps:
   a. Switch to Design view and then click the Line button in the Controls group.

b. Position the crosshairs with the line icon attached below the title in the *Form Header* section beginning a few rows of grid dots below the first letter in the title. Press and hold down the Shift key, drag to the right, release the mouse button below the last letter in the title, and then release the Shift key.

c. Click the Form Design Tools Format tab, click the Shape Outline button in the Control Formatting group, point to *Line Thickness* at the drop-down list, and then click *3 pt* (fourth option from the top).

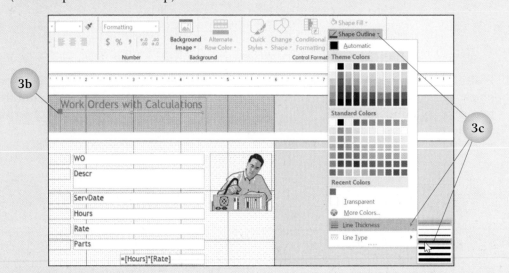

d. With the line control object still selected, click the Shape Outline button and then click the *Dark Red* color (first option in last row of *Standard Colors* section).

e. Deselect the line control object.

4. Display the form in Form view to see the line under the title.

5. Switch to Design view. If desired, adjust the length and/or position of the line.

6. Save the form.

Check Your Work

Anchoring Controls to a Form

Quick Steps

Anchor Control to Form

1. Open form in Design view.
2. Select control object(s) to be anchored.
3. Click Form Design Tools Arrange tab.
4. Click Anchoring button.
5. Click anchor position.
6. Deselect control object.
7. Save form.

 Anchoring

A control object in a form can be anchored to a section or another control object using the Anchoring button in the Position group on the Form Design Tools Arrange tab. When a control object is anchored, its position is maintained when the form is resized. For example, if an image is anchored to the top right of the *Detail* section, when the form is resized in Form view, the image automatically moves so that the relative distance between the image and the top right of the *Detail* section is maintained. If the image is not anchored and the form is resized, the position of the image relative to the edges of the form can change.

By default, *Top Left* is selected as the anchor position for each control object in a form. To change the anchor position, select the object(s), click the Form Design Tools Arrange tab, click the Anchoring button, and then click one of these options: *Stretch Down, Bottom Left, Stretch Across Top, Stretch Down and Across, Stretch Across Bottom, Top Right, Stretch Down and Right,* or *Bottom Right*. Click the option that represents how the control object is to move as the form is resized. Note that some options will cause a control object to resize as well as move when the form is changed.

1. With **4-RSRCompServ** open and the WorkOrders form open in Design view, anchor the image to the top of the *Detail* section of the form by completing the following steps:
   a. Click to select the image.
   b. Click the Form Design Tools Arrange tab.
   c. Click the Anchoring button in the Position group and then click *Top Right* at the drop-down list.

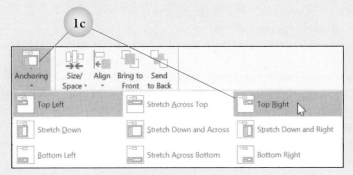

   d. Take note of the distance between the top border of the selected image and the top of the *Detail* section. ***Note: If you have a wide monitor, you may want to change the anchoring back to*** **Top Left**.
2. Display the form in Form view. Notice that the image has shifted to the right of the *Detail* section. However, it has maintained the distance between the top of the control object boundary and the top of the *Detail* section.

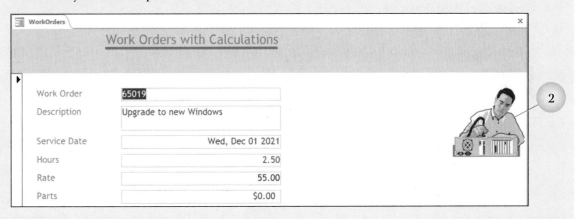

3. Save the form and then switch to Design view.

1. With **4-RSRCompServ** open and the WorkOrders form open in Design view, adjust the alignment and formatting of numeric fields by completing the following steps:
    a. Press and hold down the Shift key; click the left mouse button to select the *Rate* and *Parts* text box control objects and both calculated text box control objects; and then release the mouse button.
    b. Click the Form Design Tools Format tab, click the Align Right button in the Font group, and then click the Currency button in the Number group.
    c. Deselect the control objects.
2. Switch to Form view. Notice that the decimal point in the data in the *Hours* text box control object does not align with decimal points in the other numbers in the form. Switch back to Design view. To align the decimal point, complete the following steps:
    a. Right click the *Hours* text box control object and then click *Properties* at the shortcut menu.
    b. Click the Format tab, scroll down if necessary, and then change the Right Margin property from *0* to *0.07*.
    c. Close the Property Sheet task pane.
3. Click to select the title control object box in the *Form Header* section and then drag the bottom middle sizing handle up to decrease the height of the control object to approximately 0.4 inch on the vertical ruler. If necessary, drag down the bottom border of the *Form Header* section in order to select the sizing handle

4. Position the pointer on the top of the gray *Detail* section bar until the pointer displays as an up-and-down-pointing arrow with a horizontal line in the middle and then drag up to decrease the height of the *Form Header* section to approximately 0.5 inch on the vertical ruler.
5. Save the form.
6. Display the form in Form view and compare it with the one shown in Figure 4.12.
7. Print only the selected record with the left and right margins both set to 0.5 inch.
8. Close the form.

Check Your Work

Figure 4.12 Completed WorkOrders Form

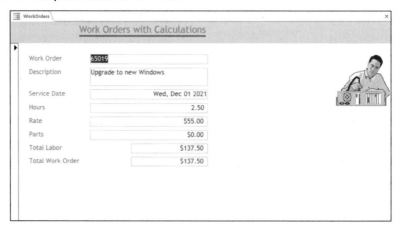

You will create a datasheet form for use in entering information into a table and set the form's properties to prevent records from being deleted.

Creating a
Datasheet Form
and Restricing
Actions

Creating a Datasheet Form and Restricting Actions

A form can be created that looks just like the datasheet of a table. With the appropriate table selected in the Navigation pane, click the Create tab, click the More Forms button in the Forms group, and then click *Datasheet* at the drop-down list. Access creates a form including all the fields from the selected table presented in a datasheet layout. The datasheet form is similar in appearance and purpose of a table datasheet, and providing the form to users instead of the underlying table prevents them from accessing and modifying the structure of the table.

To restrict what actions users can perform, display the form in Form view and use options on the Data tab of the form's Property Sheet, as shown in Figure 4.13. For example, prevent new records from being added and/or existing records from being deleted, edited, and/or filtered. Setting the *Data Entry* property to *Yes* means the user will see only a blank form when the form is opened. A data entry form is intended to be used only to add new records. The user is prevented from scrolling through existing records in the form.

Quick Steps

Create Datasheet Form
1. Select table in Navigation pane.
2. Click Create tab.
3. Click More Forms button.
4. Click *Datasheet*.
5. Save form.

 More

 datasheet

Quick Steps

Restrict Actions in Form
1. Open form in Design view.
2. Double-click Form Selector button.
3. Click Data tab.
4. Change *Allow Additions, Allow Deletions, Allow Edits,* or *Allow Filters* to *No*.
5. Close Property Sheet task pane.
6. Save form.

Figure 4.13 Form Property Sheet Task Pane with the Data Tab Selected

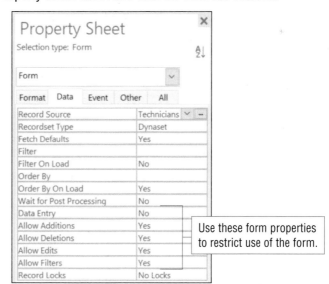

1. With **4-RSRCompServ** open, click *Technicians* in the Tables group in the Navigation pane and then click the Create tab.
2. Click the More Forms button in the Forms group and then click *Datasheet* at the drop-down list.

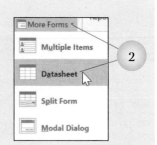

3. Review the Technicians form in the work area. Notice that the form resembles a table datasheet.
4. Switch to Design view.
5. Modify the properties of the Technicians form to prevent users from deleting records by completing the following steps:
   a. Click in a blank area to deselect the control objects.
   b. Double-click the Form Selector button at the top of the vertical ruler and to the left of the horizontal ruler to open the form's Property Sheet task pane.
   c. Click the Data tab.
   d. Double-click in the *Allow Deletions* property box to change *Yes* to *No*.
   e. Close the Property Sheet task pane.

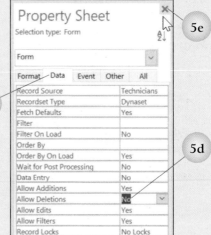

6. Click the Save button and then click OK to accept *Technicians* as the form name.
7. Click the View button arrow in the Views group on the Form Design Tools Design tab. Notice that *Datasheet View* and *Design View* are the only views available. The *Form View* option is not available at the drop-down list or in the View buttons at the right side of the Status bar.

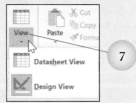

8. Click in a blank area to close the drop-down list.
9. Close the form.
10. Double-click *Technicians* in the Forms group in the Navigation pane. (Be careful to open the form object and not the table object.)
11. Click the record selector bar next to the first row in the datasheet (for technician ID 01) to select the record.
12. Click the Home tab and then look at the Delete button in the Records group. Notice that the Delete button is dimmed. This feature is unavailable because the *Allow Deletions* property box is set to *No*.

13. Best fit the width of each column, print the first page of the Technicians form in landscape orientation, and then close the form.

14. Right-click *Technicians* in the Forms group in the Navigation pane and then click *Layout View* at the shortcut menu. Notice that the datasheet form displays in a columnar layout in Layout view. The Technicians form includes a field named *Attachments*. In this field in the first record, a picture of the technician has been attached to the record. In Layout view, Access automatically opens the image file and displays the contents.

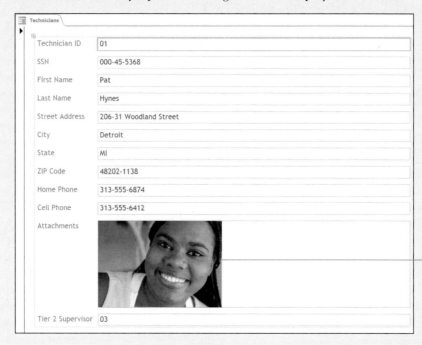

The *Attachments* field automatically displays an attached image file if one has been provided.

15. Close the form.

Check Your Work

Activity 4 Create a Blank Form with Lists   3 Parts

You will use the Blank Form tool to create a new form for maintaining the FeesSCPlans table, which is used to track service plan fees. In the form, you will create list boxes to provide an easy way to enter data for new service contract plans.

 Tutorial

Creating a Blank Form

 Blank Form

Creating a Blank Form

To quickly build a form that contains a small number of fields, use the Blank Form tool. A blank form begins with no controls or formatting and displays as a blank white page in Layout view.

To begin a new form, click the Create tab and then click the Blank Form button in the Forms group. Access opens the Field List task pane at the right side of the work area. Expand the list for the appropriate table and then add fields to the form as needed. If the Field List task pane displays with no table names, click the Show all tables hyperlink at the top of the pane.

1. With **4-RSRCompServ** open, click the Create tab and then click the Blank Form button in the Forms group.
2. If the Field List task pane at the right side of the work area does not display the table names, click the <u>Show all tables</u> hyperlink. Otherwise, proceed to Step 3.
3. Add fields from the FeesSCPlans table to the form by completing the following steps:
   a. Click the plus symbol (+) next to the FeesSCPlans table to expand the field list.
   b. Click the first field (named *ID*) in the Field List task pane and then drag the field to the top left of the form.
   c. Click the second field (named *Term*) in the Field List task pane, press and hold down the Shift key, click the last field, (named *Rate*) in the Field List task pane to select the remaining fields in the FeesSCPlans table, and then release the Shift key.
   d. Position the mouse pointer within the selected field names and then drag the group of fields to the form below the *ID* field. Release the mouse when the pink bar displays below *ID*.
   e. With the four fields added to the table still selected, press and hold down the Shift key, click the *ID* field, and then release the Shift key.
   f. Position the mouse pointer on the orange border at the right of any of the selected label control objects until the pointer changes to a left-and-right-pointing arrow and then drag the right edge of the label control objects to the right until all the label text is visible, as shown below.

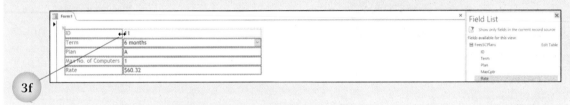

3f

4. Close the Field List task pane.
5. Save the form with the name *SCPlans*.

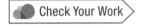 Check Your Work

Adding a List Box to a Form

A list box displays a list of values for a field within the control object. In Form view, the user can easily see the entire list for the field. Create the list of values when the control object is created or instruct Access to populate the list using values from a table or query. When a list box control object is added to the form, the List Box Wizard begins (as long as the Use Control Wizards button is toggled on). Within the List Box Wizard, specify the values to be shown in the list box.

1. With **4-RSRCompServ** open and the SCPlans form open in Layout view, add a list box control object to show the plan letters in a list by completing the following steps:
   a. Click the List Box button in the Controls group on the Form Layout Tools Design tab.
   b. Position the pointer with the list box icon attached below the *Plan* field text box control object in the form. Click the mouse when the pink bar displays between *A* and *1* in the right column. The List Box Wizard starts when the mouse is released. ***Note: If the Wizard does not start and the Use Control Wizards button is active, undo the addition, change to Design view and then back to Layout view and repeat Steps 1a and 1b.***

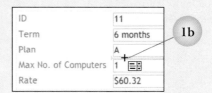

   c. At the first List Box Wizard dialog box, click *I will type in the values that I want* and then click the Next button.

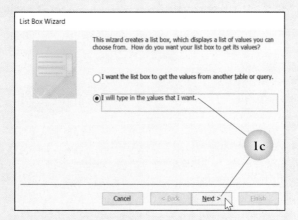

   d. At the second List Box Wizard dialog box, click in the first cell below *Col1*, type A, and then press the Tab key.
   e. Type B, press the Tab key, type C, press the Tab key, type D, and then click the Next button.

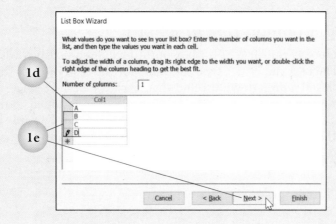

f. At the third List Box Wizard dialog box, click *Store that value in this field* option, click the option box arrow, and then click *Plan* at the drop-down list.

g. Click the Next button.

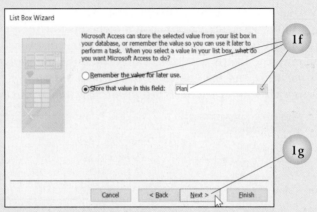

h. At the last List Box Wizard dialog box with the current text already selected in the *What label would you like for your list box?* text box, type PlanList and then click the Finish button. Access adds the list box to the form, displaying all the values entered in the list.

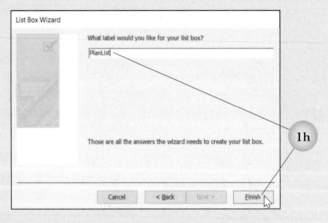

2. Save the form.

Check Your Work

Adding a Combo Box to a Form

 Tutorial

Adding a Combo Box

 Combo Box

Quick Steps

Add Combo Box
1. Open form in Layout or Design view.
2. Click Combo Box button in Controls group.
3. Click within form.
4. Create values within Combo Box Wizard.
5. Save form.

A combo box is similar to a list box but it includes a text box within the control object. The user can either type the value for the field or click the arrow to display field values in a drop-down list and then click the desired value. As happens when adding a list box, when a combo box control object is added to the form, the Combo Box Wizard begins (as long as the Use Control Wizards button is toggled on). Within the Combo Box Wizard, specify the values to be shown in the drop-down list.

1. With **4-RSRCompServ** open and the SCPlans form open in Layout view, add a combo box control object to enter the maximum number of computers in a plan by completing the following steps:

   a. Click the Combo Box button in the Controls group on the Form Layout Tools Design tab.

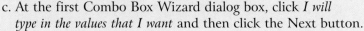

   b. Position the pointer with the combo box icon attached below the *Max No. of Computers* text box control object in the form. Click the mouse when the pink bar displays between *1* and *$60.32* in the right column. The Combo Box Wizard starts when the mouse button is released. ***Note: If the Wizard does not start and the Use Control Wizards button is active, undo the addition, change to Design view and then back to Layout view and repeat Step 1a and 1b.***

   c. At the first Combo Box Wizard dialog box, click *I will type in the values that I want* and then click the Next button.

   d. At the second Combo Box Wizard dialog box, click in the first cell below *Col1*, type *1*, and then press the Tab key.

   e. Type *2*, press the Tab key, type *3*, press the Tab key, type *4*, press the Tab key, type *5*, and then click the Next button.

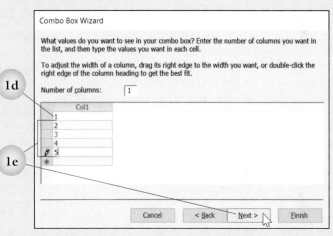

   f. At the third Combo Box Wizard dialog box, click *Store that value in this field* option, click the option box arrow, click *MaxCptr* at the drop-down list, and then click the Next button.

   g. At the last Combo Box Wizard dialog box with the current text already selected in the *What label would you like for your combo box?* text box, type CptrList and then click the Finish button. Access adds the combo box to the form, displaying a value and an option box arrow at the right side of the text box.

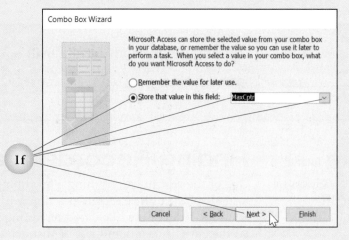

2. Double-click the label for the combo box added in Step 1g (which currently reads *CptrList*), select the text if it is not already selected, and then type Maximum Computers. Edit the label for the list box added in Activity 4b (which currently reads *PlanList*) to add a space between *Plan* and *List*.

3. Right-click the label control object above the combo box (which displays the text *Max No. of Computers*) and then click *Select Entire Row* at the shortcut menu. Press the Delete key to remove the selected row from the form.

4. Click the Title button in the Header/Footer group on the Form Layout Tools Design tab and then type the text Service Contract Plans.

5. Switch to Form view and then scroll through the records in the form.

6. Add a new record to the table by completing the following steps:
   a. Click the New button in the Records group on the Home tab.
   b. Press the Tab key to move past the *ID* field since this field is an AutoNumber data type field.
   c. Click the *Term* option box arrow and then click *2 years* at the drop-down list.
   d. Click *B* in the *Plan List* list box. Notice that *B* is entered in the *Plan* field text box control object when the letter *B* is clicked in the list box.
   e. Click the *Maximum Computers* option box arrow and then click *3* at the drop-down list.
   f. Click in the *Rate* text box and then type 360.50.

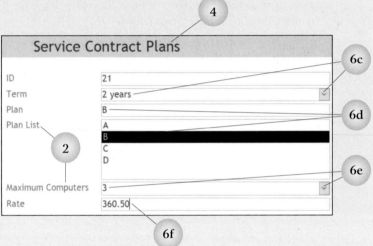

7. Print the selected record, save, and then close the form.

Check Your Work

Activity 5 Finding Records within a Form   1 Part

You will open a custom-built form and use it to find records by using a wildcard character.

Tutorial

Finding Records in a Form

Finding Records in a Form

One of the advantages of using a form for data entry and maintenance is that the form displays a single record at a time within the work area. Seeing one record at a time reduces the likelihood of editing the wrong record; the user can focus on the current record and not be distracted by other records. In a table with many records, quickly finding the specific record to be maintained or viewed is important. Use the Find feature to move to records quickly.

Quick Steps

Find Records Using Wildcard

1. Open form in Form view.
2. Click in field by which to search.
3. Click Find button or Ctrl + F.
4. Type search string, including asterisk for any variable text.
5. Click Find Next.
6. Continue clicking Find Next until search is finished.
7. Click OK.
8. Close Find and Replace dialog box.

 Find

The Find feature allows searching for records without specifying the entire field value. To do this, substitute wildcard characters in the positions for which the exact text is not specified. Two commonly used wildcard characters are the asterisk (*) and the question mark (?).

For example, suppose that a search is needed for a record by a person's last name but the correct spelling is not known. Use the asterisk wildcard character in a position for which one or more characters vary. Use the question mark wildcard character in a fixed-width word for which the records with the same number of characters in the field are to be viewed. In this case, substitute one question mark for each character not specified. Table 4.1 provides examples of how to use the asterisk and question mark wildcard characters. In Activity 5, the asterisk wildcard character will be used to locate customer records for a specified street.

Table 4.1 Examples of Using Wildcard Characters with the Find Feature

Find What Entry	In This Field	Will Find
104?	Customer ID	customer records with customer IDs that begin with *104* and have one more character, such as *1041, 1042, 1043,* and so on up to *1049*
4820?	ZIP Code	customer records with zip codes that begin with *4820* and have one more character, such as *48201, 48202,* and so on
650??	Work Order	work order records with work order numbers that begin with *650* and have two more characters, such as *65023, 65035, 65055,* and so on
313*	Home Phone	customer records with telephone numbers that begin with the *313* area code
Peter*	Last Name	customer records with last names that begin with *Peter* and have any number of characters following, such as *Peters, Peterson, Petersen, Peterovski,* and so on
4820*	ZIP Code	customer records with zip codes that begin with *4820* and have any number of characters following, such as *48201* and *48203-4841*
oak	Street Address	customer records with street addresses that have *oak* within them, such as *1755 Oak Drive, 12-234 Oak Street,* and *9 Oak Boulevard*

1. With **4-RSRCompServ** open, open the CustMaintenance form in Form view.
2. To locate the name of the customer who resides on Roselawn Street when neither the house number nor the customer's name is known, complete the following steps to find the record using a wildcard character in the criterion:
   a. Click in the *Street Address* field.
   b. Click the Find button in the Find group on the Home tab.
   c. With the insertion point positioned in the *Find What* text box, type *roselawn* and then click the Find Next button. The entry *roselawn* means "Find any record in which any number of characters before *roselawn* and any number of characters after *roselawn* exist in the active field." Access displays the first record in the form in which a match was found.

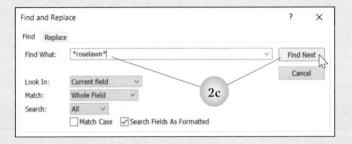

   d. Click the Find Next button to see if any other records exist for customers on Roselawn Street.
   e. At the Microsoft Access message box indicating that Access has finished searching the records, click OK.
   f. Close the Find and Replace dialog box.
3. Close the CustMaintenance form.
4. Close **4-RSRCompServ**.

Chapter Summary

- In Design view initially, a new form displays the *Detail* section, which is used to display records from the table associated with the form.

- A *Form Header* section and *Form Footer* section can be added to a form. Control objects placed in the *Form Header* section display at the top of the form or print at the beginning of a printout of records from Form view. Control objects placed in the *Form Footer* section display at the bottom of the form or print at the end of a printout of records from Form view.

- A form can contain three types of control objects: bound, unbound, and calculated.

- Click the Title button to display the *Form Header* and *Form Footer* sections and add a label control object in the *Form Header* section that contains the form name.

- Click the Label button in the Controls group to add a label control object within any section of the form.

- To specify the table to be connected to the form, use the *Record Source* property box on the Data tab of the form's Property Sheet task pane. Open this task pane by double-clicking the Form Selector button.

- Once a table has been connected to a form, click the Add Existing Fields button to open the Field List task pane.

- To add fields to a form, drag selected field names individually or in groups from the Field List task pane to the *Detail* section of the form.

- Use the move handle (the large dark-gray square at the top left of the selected control object) to move a selected control object independently of its connected label control or text box control object.

- Use the buttons in the Font group on the Form Design Tools Format tab to apply various types of formatting to selected control objects.

- Format multiple control objects at the same time by drawing a selection rectangle around a group of adjacent control objects or pressing and holding down the Shift key while clicking individual control objects.

- Open the Tab Order dialog box to change the order that fields are selected in when using the Tab key to move from field to field in Form view.

- Add a tab control object to a form to organize groups of related fields or insert subforms that display on separate pages.

- Click the Subform/Subreport button in the Controls group to add a subform to a page.

- Add a calculated control object to a form using the Text Box button in the Controls group.

- Type a formula in a text box control object (which displays *Unbound*) beginning with an equals sign. Enclose field names within the formula in square brackets.

- Use the Size/Space button and Align button in the Sizing & Ordering group on the Form Design Tools Arrange tab to position, align, size, or adjust spacing between multiple selected control objects.

- Add graphics to a form using the Logo button in the Header/Footer group or the Insert Image button in the Controls group.

- Draw a horizontal or vertical line in a form using the Line button in the Controls group. To draw a straight line, press and hold down the Shift key while dragging. Use the Shape Outline button on the Form Design Tools Format tab to adjust a line's thickness, type, or color.

- Online images can be added to a form after being inserted in a Microsoft Word document and then saved to a folder using standard Windows commands. Insert the saved image in the form using the Insert Image button.

- Change a control object's Size Mode property to resize an image within a control object while maintaining the original proportions of height and width.

- Use the Anchoring button in the Position group to anchor a control object to a section or another control object so that its position relative to the edges of the form is maintained when the form is resized.

- A datasheet form looks like a table datasheet.

- Use options on the Data tab of the form's Property Sheet task pane to restrict what actions a user can perform when viewing records in Form view.

- Use the Blank Form tool in the Forms group on the Create tab to create a new form with no controls or formatting applied. The form displays as a blank white page in Layout view with the Field List task pane open at the right of the work area.

- A list box displays a list of values for a field within the control object. The values to be shown in the list box can be specified within the List Box Wizard.
- A combo box includes a text box within the control object so the user can type the field value into the text box or click the option box arrow to pick the field value from a drop-down list. The values to be shown in the drop-down list can be specified within the Combo Box Wizard.
- Find specific records quickly using the Find feature.
- The Find feature allows the use of wildcard characters, such as the asterisk and question mark, to search records without specifying the entire field value.
- To find a record in a form, click in a field, click the Find button, and then enter the search criterion in the *Find What* text box.

Commands Review

FEATURE	RIBBON TAB, GROUP	BUTTON, OPTION	KEYBOARD SHORTCUT
add existing fields	Form Design Tools Design, Tools		
adjust size of multiple controls	Form Design Tools Arrange, Sizing & Ordering		
align multiple controls at same position	Form Design Tools Arrange, Sizing & Ordering		
anchor controls to form	Form Design Tools Arrange, Position		
blank form	Create, Forms		
change tab order of fields	Form Design Tools Design, Tools		
combo box	Form Layout Tools Design, Controls		
create form in Design view	Create, Forms		
datasheet form	Create, Forms		
Design view	Home, Views OR Form Design Tools Design, Views		
equal spacing between control objects	Form Design Tools Arrange, Sizing & Ordering		
Find	Home, Find		Ctrl + F
Form view	Home, Views OR Form Design Tools Design, Views		
insert image	Form Design Tools Design, Controls		
label control object	Form Design Tools Design, Controls	Aa	
line	Form Design Tools Design, Controls		
list box	Form Design Tools Design, Controls		

FEATURE	RIBBON TAB, GROUP	BUTTON, OPTION	KEYBOARD SHORTCUT
Property Sheet task pane	Form Design Tools Design, Tools		F4
subform	Form Design Tools Design, Controls		
tab control object	Form Design Tools Design, Controls		
text box control object	Form Design Tools Design, Controls		
title control object	Form Design Tools Design, Header/Footer		

Microsoft Access® Level 2

Unit 2

Advanced Reports, Access Tools, and Customizing Access

Microsoft®
Access®

Creating and Using Custom Reports

Performance Objectives

Upon successful completion of Chapter 5, you will be able to:

1 Create a custom report in Design view using all five report sections

2 Connect a table or query to a report and add fields

3 Move, size, format, and align control objects

4 Insert a subreport into a report

5 Add page numbers and date and time control objects to a report

6 Add graphics to a report

7 Group records and add functions and totals

8 Add a calculated field to a report

9 Modify section or group properties to control print options

10 Insert and edit a chart in a report

11 Create a blank report

12 Add tab control objects, list boxes, combo boxes, and hyperlinks to a report

13 Change the shape of a control object

14 Change the tab order of fields

Reports are used to generate printouts from the tables in a database. Although the Print feature can be used to print data from a table datasheet, query results datasheet, or form, the formatting of the data in this printout cannot be changed or customized. The Report feature provides tools and options that can be used to control the content and formatting to produce professional-quality reports that serve particular purposes. In this chapter, you will learn how to build custom reports.

 Data Files

Before beginning chapter work, copy the AL2C5 folder to your storage medium and then make AL2C5 the active folder.

The online course includes additional training and assessment resources.

Activity 1 **Design and Create a Custom Report** **5 Parts**

You will create a custom report in Design view with fields from two tables and insert a subreport.

Tutorial

Creating Custom
Reports Using
Design View

Report
Design

Quick Steps

**Start New Report
in Design View**
1. Click Create tab.
2. Click Report Design
   button.

Add Report Title
1. Open report in
   Design view.
2. Click Title button.
3. Type title text.
4. Press Enter key.

**Add Label
Control Object**
1. Open report in
   Design view.
2. Click Label button.
3. Drag to create
   control object.
4. Type label text.
5. Press Enter key.

Creating Custom Reports Using Design View

Access provides the Report tool and Report Wizard to help quickly create reports that can be modified later. Customize the content, format, and layout of a report in Layout view or Design view. In most cases, use one of the report tools to generate the report structure and then customize the report using a different view. However, if a report requires several custom options, begin in Design view and build the report from scratch. Click the Create tab and then click the Report Design button in the Reports group to create a new report using Design view, as shown in Figure 5.1.

Creating a custom report in Design view involves using the same techniques taught in Chapter 4 for designing and building a custom form. Adding a title; connecting a table or query to the report; adding fields; and aligning, moving, resizing, and formatting control objects is done the same way as customizing a form.

A report can contain up to five sections, each of which is described in Table 5.1. *Group Header* and *Group Footer* sections can be added to group records that contain repeating values in a field, such as a department or city. How to use these additional sections will be demonstrated in Activity 3. A report that is grouped by more than one field can have multiple *Group Header* and *Group Footer* sections.

Figure 5.1 Report Design View

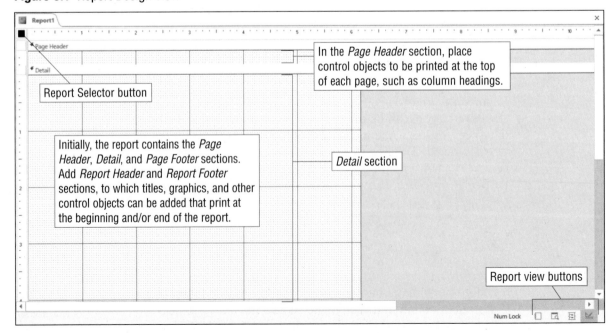

Table 5.1 Report Sections

Report Section	Description
Report Header	Content prints at the beginning of the report. Add control objects to this section for the report title and company logo or another image.
Page Header	Content prints at the top of each page in the report. Add control objects to this section for column headings in a tabular report format.
Detail	Add control objects to this section for the fields from the table or query that make up the body of the report.
Page Footer	Content prints at the bottom of each page in the report. Add a control object to this section to print a page number at the bottom of each page.
Report Footer	Content prints at the end of the report. Add a control object in this section to print a grand total or perform another function, such as calculating the average, maximum, minimum, or count.

Activity 1a Starting a New Report Using Design View and Adding a Title and Label Control Object

1. Open **5-RSRCompServ** and enable the content.
2. Click the Create tab and then click the Report Design button in the Reports group.
3. Add a title in the *Report Header* section of the report by completing the following steps:
   a. With the Report Design Tools Design tab active, click the Title button in the Header / Footer group. Access adds the *Report Header* section above the *Page Header* section and places a title object containing the selected text *Report1*.
   b. Type RSR Computer Service Work Orders and then press the Enter key.

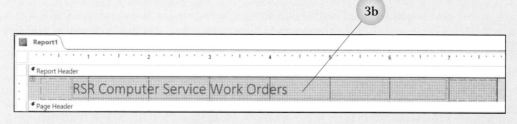

4. With the title control object still selected, click the Report Design Tools Format tab and then click the Center button in the Font group.
5. Drag the right edge of the report grid to the right until it is aligned at the 8-inch position on the horizontal ruler.
6. Scroll down the report until the *Page Footer* and *Report Footer* sections are visible. The *Report Footer* section was added to the design grid at the same time the *Report Header* section was added when the title was created in Step 3.

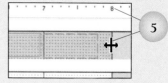

7. Drag down the bottom edge of the *Report Footer* section until the bottom of the report is aligned at the 0.5-inch position on the vertical ruler.
8. Click the Report Design Tools Design tab and then click the Label button in the Controls group. Add a label control object containing your first and last names at the left side of the *Report Footer* section.
9. Click in a blank area of the report to deselect the label control object.
10. Save the report and name it *WorkOrders*.

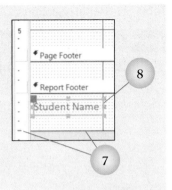

 Check Your Work

 Tutorial

Connecting a Table to a Report and Adding Fields

Quick Steps

Connect Table or Query to Report
1. Open report in Design view.
2. Double-click Report Selector button.
3. Click Data tab in Property Sheet task pane.
4. Click arrow in *Record Source* property box.
5. Click table or query.
6. Close Property Sheet task pane.

Add Fields to Report
1. Click Add Existing Fields button.
2. Drag field name(s) from Field List task pane to *Detail* section.

Add Field from Related Table
1. Open Field List task pane.
2. Click Show all tables hyperlink.
3. Click expand button next to table name in *Fields available in related tables* section.
4. Drag field name from related table list to *Detail* section.

 Report Selector

 Add Existing Fields

Connecting a Table or Query to a Report and Adding Fields

A new report that is started in Design view does not have a table or query associated with it. To display data in the report, Access needs to know what source to draw the data from. Connect a table or query to the report using the Record Source property in the report's Property Sheet task pane. Make sure to complete this step before adding fields to the *Detail* section.

The steps for connecting a table or query to a report are the same as the steps for connecting a table to a form. Double-click the Report Selector button above the vertical ruler and left of the horizontal ruler to open the report's Property Sheet task pane. Click the Data tab and then select the table or query name in the drop-down list at the *Record Source* property box. Display the Field List task pane and then drag individual fields or a group of fields from the table or query to the *Detail* section. After adding the fields, move and resize the control objects as needed.

The Field List task pane displays in one of two ways: showing one section only, with the fields from the table or query associated with the report, or showing two additional sections, with fields from other tables in the database. If the Field List task pane contains only the fields from the associated table or query, add fields from other tables by displaying other table names in the database within the Field List task pane.

At the top of the Field List task pane, Access displays a Show all tables hyperlink. Click the hyperlink to display the *Fields available in related tables* and *Fields available in other tables* sections. Next to each table name is an expand button, which displays as a plus symbol. Click the expand button next to the table name to display the fields stored within the table and then drag the field name(s) to the *Detail* section of the report. You will perform these steps in Activity 1b.

1. With **5-RSRCompServ** open and the WorkOrders report open in Design view, scroll to the top of the report in the work area.
2. Connect the WorkOrders table to the report so that Access knows which fields to display in the Field List task pane by completing the following steps:
   a. Double-click the Report Selector button at the top of the vertical ruler and left of the horizontal ruler to open the report's Property Sheet task pane.
   b. Click the Data tab in the Property Sheet task pane, click the *Record Source* property option box arrow, and then click *WorkOrders* at the drop-down list.
   c. Close the Property Sheet task pane.

3. Click the Add Existing Fields button in the Tools group on the Report Design Tools Design tab to open the Field List task pane.
4. Add fields from the WorkOrders table and related fields from the Customers table by completing the following steps:
   a. Click the Show all tables hyperlink at the top of the Field List task pane. Access adds two sections to the pane: one containing related tables and the other containing tables in the database that do not have established relationships with the report's table. Next to each table name is an expand button (plus symbol), which is used to display field names for the table. *Note: Skip this step if the **Show only fields in the current record source** hyperlink displays at the top of the Field List task pane. This means that the other sections have already been added to the pane.*

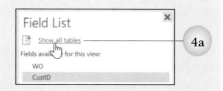

   b. Click the expand button next to *Customers* in the *Fields available in related tables* section of the Field List task pane. Access expands the list to display the field names in the Customers table.

   c. Drag the *WO, CustID, WODate,* and *Descr* fields from the WorkOrders table to the design grid, as shown below.
   d. Drag the *FName* and *LName* fields from the Customers table to the design grid, as shown at the right. Notice that the Customers table and field names move to the *Fields available for this view* section in the Field List task pane after the first field has been added from the Customers table to the *Detail* section.

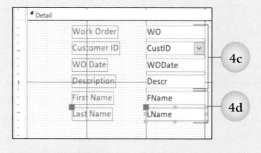

5. Close the Field List task pane.
6. Save the report.

Tutorial

Moving Control
Objects to Another
Section

Moving Control Objects to Another Section

When a field is added to the *Detail* section of a report, a label control object containing the caption or field name is placed to the left of a text box control object that displays the field value from the record when the report is viewed or printed. Recall from Chapter 4 that the same thing happens when a form is customized.

In the WorkOrders report, the label control object for each field needs to be moved to the *Page Header* section so that the field names or captions print at the top of each page as column headings. In Activity 1c, the control objects will be cut from the *Detail* section and pasted into the *Page Header* section.

Quick Steps

Move Control Objects to Another Section
1. Open report in Design view.
2. Select control objects to be moved.
3. Click Home tab.
4. Click Cut button.
5. Click bar of section to move control objects to.
6. Click Paste button.
7. Deselect controls.

Activity 1c Moving Control Objects to Another Section

Part 3 of 5

1. With **5-RSRCompServ** open and the WorkOrders report open in Design view, move the label control objects from the *Detail* section to the *Page Header* section by completing the following steps:
   a. Click to select the *Work Order* label control object.
   b. Press and hold down the Shift key, click to select each of the other label control objects, and then release the Shift key.
   c. Click the Home tab and then click the Cut button in the Clipboard group.
   d. Click the *Page Header* section bar.
   e. Click the Paste button in the Clipboard group. (Do not click the button arrow.) Access pastes the label control objects and expands the *Page Header* section.
   f. Deselect the control objects.
2. Click to select the *Customer ID* label control object and then move the control object to the top of the *Page Header* section next to the *Work Order* label control object by hovering the mouse pointer over the *Customer ID* label control object until the four-headed arrow move icon displays and then dragging the control object to the location shown below.

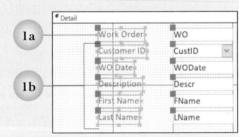

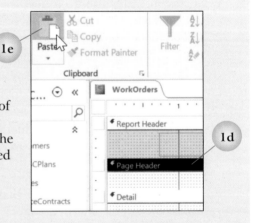

3. Move the remaining four label control objects to the top of the *Page Header* section in the order shown in the image below by completing a step similar to Step 2.
4. Drag the top of the *Detail* section bar up until the top of the bar is aligned at the bottom edge of the label control objects in the *Page Header* section, as shown below.

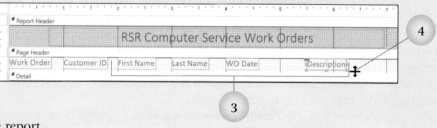

5. Save the report.

Check Your Work

Tutorial

Formatting a
Report

Themes

Applying a Theme

Apply a theme to a report using the Themes button on the Report Design Tools Design tab. The theme sets the default colors and fonts for the report. The themes available in Access align with the themes available in Word, Excel, and PowerPoint. This allows having the same look in Access reports as in other documents, worksheets, and presentations. Note that changing a theme for one report in a database automatically changes the theme for all the reports and forms in the database.

Activity 1d Moving Controls, Resizing Controls, and Applying a Theme   Part 4 of 5

1. With **5-RSRCompServ** open and the WorkOrders report open in Design view, move each text box control object in the *Detail* section below its associated label control object in the *Page Header* section so that the field values align below the correct column headings in the report, as shown below.

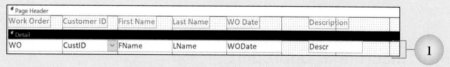

2. Click the Report Design Tools Design tab, click the View button arrow in the Views group, and then click *Print Preview* at the drop-down list. (Note that there is also a Print Preview button in the view area at the right side of the Status bar.) Notice that the field value in the *WO Date* column displays pound symbols (#); this indicates that the field's text box control object needs to be widened. **Note: If you receive an error message with the text The section width is greater than the page width, click OK.**

3. Return to Design view by clicking the Design View button in the view area at the bottom right side of the Status bar, next to the Zoom slider.

4. Resize the *WODate* text box control object in the *Detail* section so that the right edge of the control meets the left edge of the *Descr* text box control object.
5. Resize the *Descr* text box control in the *Detail* section so that the right edge of the control object is aligned at approximately the 7.5-inch position on the horizontal ruler.
6. Deselect the *Descr* text box control object.
7. Click the Themes button in the Themes group on the Report Design Tools Design tab and then click *Organic* at the drop-down gallery (third column, second row in the *Office* section).

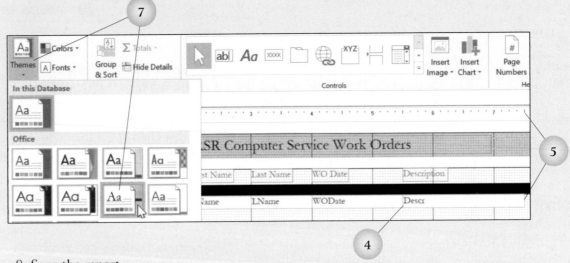

8. Save the report.
9. Display the report in Print Preview to review the changes made in this activity. Switch back to Design view when finished.

Check Your Work

Tutorial

Inserting a Subreport

Inserting a
Subreport

A report that is inserted inside another report is called a *subreport*. Similar to a nested query, a subreport allows for reusing a group of fields, formats, and calculations in more than one report without having to recreate the setup each time.

Subform/
Subreport

Use the Subform/Subreport button in the Controls group on the Report Design Tools Design tab to insert a subreport into a report. The report into which the subreport is inserted is called the *main report*. Adding a related table or query as a subreport creates a control object within the main report that can be moved, formatted, and resized independently of the other control objects. Make sure the Use Control Wizards button is toggled on in the expanded Controls group before clicking the Subform/Subreport button. This will enable the subreport to be added using the SubReport Wizard, shown in Figure 5.2.

Quick Steps

Insert Subreport
1. Open report in Design view.
2. Make sure *Use Control Wizards* is active.
3. Click Subform/ Subreport button.
4. Drag crosshairs to appropriate height and width in *Detail* section.
5. Click Next.
6. Choose table or query and fields.
7. Click Next.
8. Choose field by which to link main report with subreport.
9. Click Next.
10. Click Finish.

Figure 5.2 First Dialog Box in the SubReport Wizard

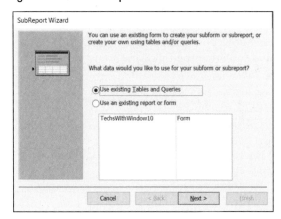

A subreport is stored as a separate object outside the main report. An additional report name will display in the Navigation pane with *subreport* at the end. Do not delete a subreport object in the Navigation pane. If the subreport object is deleted, the main report will no longer display the fields from the related table or query in the report.

Activity 1e Inserting a Subreport

Part 5 of 5

1. With **5-RSRCompServ** open and the WorkOrders report open in Design view, insert a subreport into the WorkOrders report with fields from a query for the service date, labor, and parts for each work order by completing the following steps:

   a. By default, the Use Control Wizards button is toggled on in the Controls group. Click the More button in the Controls group to expand the Controls group and view the buttons and the Controls drop-down list. View the current status of the Use Control Wizards button; if it displays with a gray background, the feature is active. If the button is gray, click in a blank area to close the expanded Controls list. If the feature is not active (displays with a white background), click the Use Control Wizards button to turn on the feature.

   b. Click the More button in the Controls group and then click the Subform/Subreport button.

   c. Move the crosshairs with the subreport icon attached to the *Detail* section below the *WO* text box control object and then drag down and to the right to create a subreport object of the approximate height and width shown below. Release the mouse and the SubReport Wizard begins.

A gray background means the Use Control Wizards button is toggled on. Check the status and click the button to turn on the feature if necessary in Step 1a.

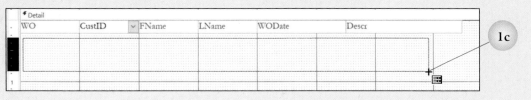

d. With *Use existing Tables and Queries* already selected at the first SubReport Wizard dialog box, click the Next button.

e. At the second SubReport Wizard dialog box, select the query fields to be displayed in the subreport by completing the following steps:

1) Click the *Tables/Queries* option box arrow and then click *Query: TotalWorkOrders* at the drop-down list.

2) Move all the fields from the *Available Fields* list box to the *Selected Fields* list box.

3) Click the Next button.

f. At the third SubReport Wizard dialog box, choose the field by which to link the main report with the subreport by completing the following steps:

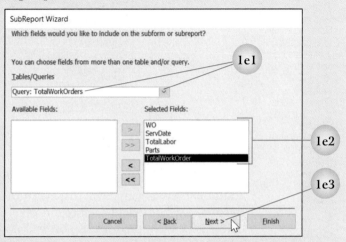

1) With *Choose from a list* and the first option in the list box selected, read the description of the linked field below the list box. The text indicates that the main report will be linked to the subreport using the *CustID* field. This is not the correct field; your report is to show the service date, labor, and parts based on the work order number.

2) Click the second option in the list box and then read the text below the list box.

3) Since the second option indicates that the two reports will be linked using the *WO* field, click the Next button.

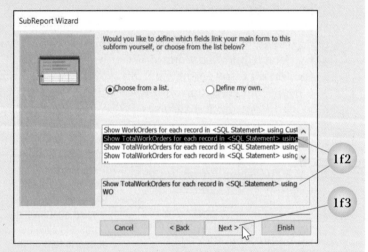

g. Click the Finish button at the last SubReport Wizard dialog box to accept the default subreport name *TotalWorkOrders subreport*.

2. Access inserts the subreport with a label control object above it that contains the name of the subreport. Click the label control displaying the text *TotalWorkOrders subreport* to select the object and then press the Delete key.

3. Click the Report View button in the Views group on the Report Design Tools Design tab. (Note that a Report View button is also included in the View group, the first button on the right side of the Status bar.) Report view is not the same as Print Preview. Report view displays the report with data in the fields and is useful for viewing reports within the database. However, this view does not show how the report will look when printed. For printing purposes, always use Print Preview to resize and adjust control objects.

4. Notice that the work order number in the subreport is the same work order number that is displayed in the first record in the main report.

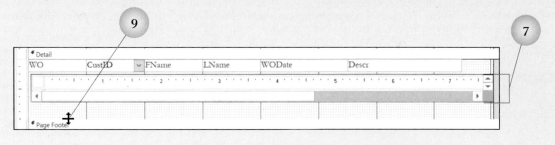

5. Switch back to Design view.
6. Now that you know the subreport is linked correctly to the main report, the work order number does not need to be displayed in the subreport. Delete the work order number control objects in the subreport by completing the following steps:
   a. Click to select the subreport control object and then drag down the bottom middle sizing handle to increase the height of the subreport until all the controls are visible in the *Report Header* and *Detail* sections.
   b. Click to select the *Work Order* label control object in the *Report Header* section, press and hold down the Shift key, click the *WO* text box control object in the *Detail* section in the subreport, and then release the Shift key.
   c. Press the Delete key.
   d. Select all the label control objects in the *Report Header* section and all the text box control objects in the *Detail* section. Move the control objects to the left so the left edge of the *ServDate* control objects are at the left margin. Click outside the subreport to deselect the control objects.
7. Click to select the subreport control object and then drag up the bottom middle sizing handle of the control until the height of the subreport is approximately 0.5 inch.
8. Scroll down the report until the *Page Footer* section bar is visible.
9. Drag up the top of the *Page Footer* section bar until the section bar is just below the subreport control object at the 1-inch mark in the *Detail* section, as shown below.

10. Save the report and then switch to Report view to view the revised report. Resizing the *Detail* section in Step 9 reduces the space between sections and allows more records and related subreport records to display on the page.

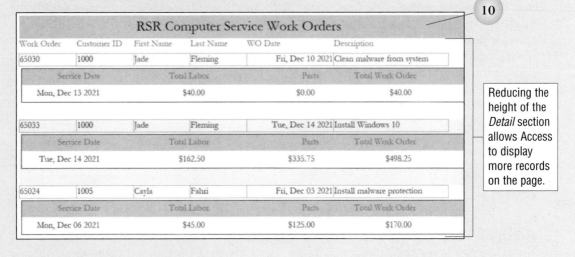

Reducing the height of the *Detail* section allows Access to display more records on the page.

11. Close the report.

Check Your Work

Activity 2 Add Features and Enhance a Report   2 Parts

You will modify the WorkOrders report to add page numbers, date and time control objects, and graphics.

 Tutorial

Adding Page Numbers and the Date and Time Control Objects to a Report

Adding Page Numbers and the Date and Time Control Objects to a Report

When creating a report using the Report tool, page numbers and the current date and time are automatically added to the top right of the report. The Report Wizard automatically inserts the current date at the bottom left and page numbers at the bottom right of the report.

In Design view, add page numbers to a report using the Page Numbers button in the Header / Footer group on the Report Design Tools Design tab. Click the button to open the Page Numbers dialog box, shown in Figure 5.3. Choose the appropriate format, position, and alignment for the page number and then click OK. Access inserts a control object in the *Page Header* or *Page Footer* section, depending on the *Position* option selected in the dialog box. Including page numbers in a report allows referring to specific pages and putting pages back in order if they get rearranged.

Quick Steps

Add Page Numbers
1. Open Report in Design view.
2. Click Page Numbers button.
3. Select format, position, and alignment options.
4. Click OK.

 Page Numbers

Quick Steps

Add Date and/or Time
1. Open report in Design view.
2. Click Date and Time button.
3. Select date and/or time options.
4. Click OK.

Date and Time

Add the current date and/or time in the *Report Header* section by clicking the Date and Time button in the Header / Footer group to open the Date and Time dialog box, shown in Figure 5.4. By default, both the *Include Date* and *Include Time* check boxes contain check marks. Access creates one control object for the date format and another for the time format. Access places the date control object above the time control object and aligns both at the right edge of the *Report Header* section. Once the control objects have been inserted, they can be moved to another section in the report. Adding a date and/or time control object means that the current date and/or time the report is printed will be included on the printout. At a minimum, always include a date control. Depending on users' needs, the time control may or may not be necessary.

Figure 5.3 Page Numbers Dialog Box

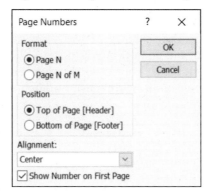

Figure 5.4 Date and Time Dialog Box

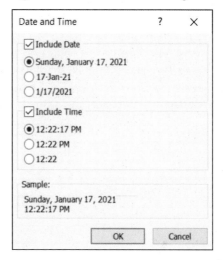

Activity 2a Adding Page Numbers and the Date and Time to a Report   Part 1 of 2

1. With **5-RSRCompServ** open, right-click *WorkOrders* in the Report group in the Navigation pane and then click *Design View* at the shortcut menu.
2. When the subreport was inserted in Activity 1e, the width of the report may have been automatically extended beyond the page width. Look at the Report Selector button.
   If a green diagonal triangle displays in the upper left corner, correct the page width by completing the following steps. (Skip this step if the Report Selector button does not display with a green diagonal triangle.)
   a. Click the subreport control object to display the orange border and sizing handles. Point to the orange border and then drag the subreport left until the left edge is at the left edge of the *Detail* section.
   b. Drag the right middle sizing handle left to decrease the subreport width until the right edge of the subreport is aligned with the right edge of the *Descr* text box control object above it.

c. Click the green triangle to display the Error-Checking Options button and then click the button to display the drop-down list of options.

d. Click *Remove Extra Report Space* at the drop-down list to automatically decrease the width of the report. Notice that the green diagonal triangle is removed from the Report Selector button once the report width has been corrected.

3. Add a page number at the bottom center of each page by completing the following steps:

a. Click the Page Numbers button in the Header / Footer group on the Report Design Tools Design tab.

b. Click *Page N of M* in the *Format* section of the Page Numbers dialog box.

c. Click *Bottom of Page [Footer]* in the *Position* section.

d. With the *Alignment* option box set to *Center* and a check mark in the *Show Number on First Page* check box, click OK. Access adds a control object in the center of the *Page Footer* section that contains the codes required to print the page numbers centered at the bottom of all pages, including the first page.

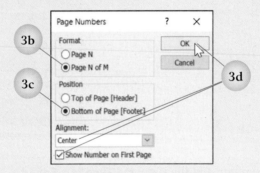

4. Add the current date and time to the end of the report along with a label control object that contains the text *Date and Time Printed:* by completing the following steps:

a. Click the Date and Time button in the Header / Footer group on the Report Design Tools Design tab.

b. Click the second option in the *Include Date* section in the Date and Time dialog box, which displays the date in the format *dd-mmm-yy*—for example, *17-Jan-21*.

c. Click the second option in the *Include Time* section, which displays the time in the format *hh:mm AM/PM*—for example, *2:05 PM*.

d. Click OK. Access adds two control objects, one above the other, at the right side of the *Report Header* section with the date code *=Date()* and time code *=Time()*.

e. Select the date and time control objects added to the *Report Header* section. Click the Home tab and then click the Cut button.

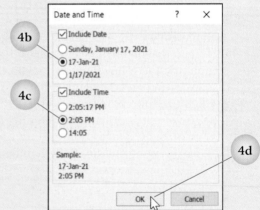

f. Click the *Report Footer* section bar and then click the Paste button. Access pastes the two control objects at the left side of the *Report Footer* section. With the date and time control objects still selected, position the mouse pointer on the orange border until the pointer displays with the four-headed arrow move icon and then drag the control objects to the right side of the *Report Footer* section, aligning the right edge of the control objects near the right edge of the report grid.

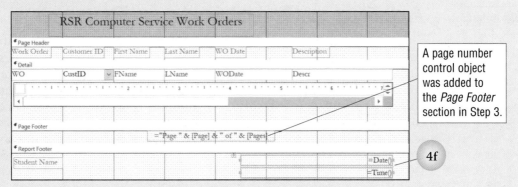

A page number control object was added to the *Page Footer* section in Step 3.

4f

g. Resize the date and time controls to be approximately 0.75-inch wide as shown at the right.

h. Create the two label control objects, type the text Date Printed: and Time Printed:, and then position the label control objects left of the date control object and time control object, as shown at the right.

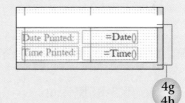

4g
4h

5. Save the report and then display it in Print Preview.

6. Scroll to the bottom of the first page to view the page number.

7. Click the Last Page button in the Page Navigation bar to scroll to the last page in the report and view the date and time.

8. Notice that in Print Preview, the subreport data is cut off at the right edge of the report, which means the total work order value is not visible. Exit Print Preview to switch back to Design view.

9. Adjust the size and placement of the subreport control objects by completing the following steps:

a. Since the subreport control objects are not visible within the WorkOrders report, it is easier to work within the separate TotalWorkOrders subreport to make changes to its content. Close the WorkOrders report.

b. Right-click *TotalWorkOrders subreport* in the Navigation pane and then click *Design View* at the shortcut menu.

c. Using the diagram below as a guide, adjust the widths of the Total Labor and/or Parts label control objects and text box control objects. Remove any space between the control objects.

d. Drag the right edge of the grid left to approximately the 7-inch position on the horizontal ruler.

9c

9d

10. Save and close the TotalWorkOrders subreport.
11. Open the WorkOrders report. Review the data.
12. Display the report in Design view.

 Check Your Work

Tutorial

Adding a Graphic to
a Report

 Insert Image

Adding a Graphic to a Report

The same techniques learned in Chapter 4 for adding online images or drawing lines in a form in Design view can be applied to a report. An image in a standard picture file format (such as .gif, .jpg, or .png) can be inserted in an image control object. Click the Insert Image button in the Controls group on the Report Design Tools Design tab, browse to the drive and folder containing the image, double-click the image file name, and then drag to create an image control object of the required height and width within the report. Recall from Chapter 4 that when a control object containing an image is resized, parts of the image can be cut off. Display the Property Sheet task pane for the control object and then change the Size Mode property to *Zoom* or *Stretch* to resize the image to the height and width of the control object.

Activity 2b Adding Graphics and Formatting Control Objects   Part 2 of 2

1. With **5-RSRCompServ** open and the WorkOrders report open in Design view, insert a company logo in the report by completing the following steps:
   a. Position the mouse pointer on the top of the *Page Header* section bar until the pointer displays as an up-and-down-pointing arrow with a horizontal line in the middle and then drag down approximately 0.25 inch to increase the height of the *Report Header* section.
   b. Click the Insert Image button in the Controls group on the Report Design Tools Design tab and then click *Browse* at the drop-down list.
   c. At the Insert Picture dialog box, navigate to the drive and/or folder for the AL2C5 data files on your storage medium and then double-click the file named ***RSRlogo***.
   d. Position the crosshairs with the image icon attached at the top of the *Report Header* section near the 6-inch position on the horizontal ruler and then drag to create an image control object of the approximate height and width shown.

2. Select the title control object in the *Report Header* section and then click the Bold button in the Font group on the Report Design Tools Format tab.

3. Draw and format a horizontal line below the title by completing the following steps:
   a. Click the More button in the Controls group on the Report Design Tools Design tab and then click the Line button in the expanded Controls group.
   b. Position the crosshairs with the line icon attached below the first letter in the title in the *Report Header* section, press and hold down the Shift key, click the left mouse button and drag to the right, release the mouse button below the last letter in the title, and then release the Shift key.
   c. Click the Report Design Tools Format tab and then click the Shape Outline button in the Control Formatting group.
   d. Point to *Line Thickness* and then click *3 pt* (fourth option).
   e. With the line still selected, click the Shape Outline button, click *Blue-Gray, Accent 3* (seventh column, first row in the *Theme Colors* section), and then deselect the line.

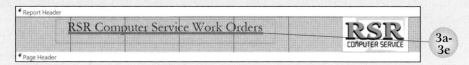

4. Draw and format a horizontal line below the column headings by completing the following steps:
   a. Drag down the top of the *Detail* section bar approximately 0.25 inch to add grid space in the *Page Header* section.
   b. Click the Report Design Tools Design tab, click the More button in the Controls group, and then click the Line button in the expanded Controls group.
   c. Draw a straight horizontal line that extends the width of the report along the bottom of the label control objects in the *Page Header* section.
   d. Click the Shape Outline button on the Report Design Tools Format tab, point to *Line Thickness*, and then click *2 pt*. Change the line color to the same Blue-Gray applied to the line below the report title.
   e. Deselect the line.

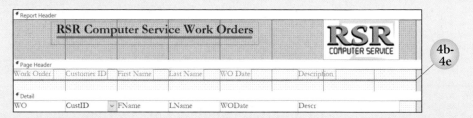

5. Format, move, and resize the control objects as follows:
   a. Select all the label control objects in the *Page Header* section, apply bold formatting, change the font size to 12 points, and then apply the Blue-Gray, Accent 3, Darker 50% font color (seventh column, last row in the *Theme Colors* section).
   b. Resize the control objects as needed to show all the label text after increasing the font size.
   c. Move the *WO Date* label control object in the *Page Header* section until the left edge of the control object is aligned at the 4.5-inch position on the horizontal ruler.
   d. Click the *Report Header* section bar, click the Shape Fill button in the Control Formatting group on the Report Design Tools Format tab, and then click *White, Background 1* (first column, first row in the *Theme Colors* section).
   e. Select all the text box control objects in the *Detail* section. Open the Property Sheet task pane, click the Format tab, click in the *Border Style* property box, click the arrow that appears, and then click *Transparent* at the drop-down list. This removes the borders around the data in the fields.

f. With all the text box control objects still selected, click the *Fore Color* property box on the Format tab of the Property Sheet task pane. Click the Build button to open the color palette and then click *Black, Text 1* (second column, first row in the *Theme Colors* section).

g. Deselect the text box control objects and then select the *WODate* text box control object. Click in the *Width* property box in the Property Sheet task pane, change *1.5* to *1.4*.

h. Select the *Descr* text box control object, change the number in the *Width* property box to *2.25*, and then close the Property Sheet task pane.

6. Display the report in Report view.

7. Compare the report with the partial report shown in Figure 5.5. If necessary, return to Design view; adjust the formats, alignments, or positions of the control objects; and then redisplay the report in Report view.

8. Save the report.

9. Print the report. ***Note: This report is four pages long. Check with your instructor before printing it.***

10. Close the WorkOrders report.

Check Your Work

Figure 5.5 Partial View of the Completed WorkOrders Report

RSR Computer Service Work Orders					RSR COMPUTER SERVICE
Work Order	Customer ID	First Name	Last Name	WO Date	Description
65030	1000	Jade	Fleming	Fri, Dec 10 2021	Clean malware from system
Service Date	Total Labor		Parts	Total Work Order	
Mon, Dec 13 2021	$40.00		$0.00	$40.00	
65033	1000	Jade	Fleming	Tue, Dec 14 2021	Install Windows 10
Service Date	Total Labor		Parts	Total Work Order	
Tue, Dec 14 2021	$162.50		$335.75	$498.25	
65024	1005	Cayla	Fahri	Fri, Dec 03 2021	Install malware protection
Service Date	Total Labor		Parts	Total Work Order	
Mon, Dec 06 2021	$45.00		$125.00	$170.00	

Activity 3 Group Records and Add Functions to Count and Sum   3 Parts

You will create a new report using the Report Wizard and then modify the report in Design view to add count and sum functions.

Tutorial

Review: Creating a Report Using the Report Wizard

Tutorial

Grouping Records in a Report

Grouping Records and Adding Functions and a Calculated Field in a Report

A field that contains repeated values—such as a department, city, or name—is an appropriate field to group records on in a report. For example, organize a report to show all the records for the same department or city. By summarizing the records by a common field value, totals can be produced using functions to calculate the

Figure 5.6 Example Report with the Work Order Records Grouped by Customer

RSR COMPUTER SERVICE	Work Orders by Customer					23-Jan-22 2:49 PM
Customer ID	First Name	Last Name	Work Order	Description	Service Date	Total Work Order
1000	Jade	Fleming	65030	Clean malware from system	Mon, Dec 13 2021	$40.00
			65033	Install Windows 10	Tue, Dec 14 2021	$498.25
					Customer Total	$538.25
1005	Cayla	Fahri	65024	Install malware protection	Mon, Dec 06 2021	$170.00
					Customer Total	$170.00
1008	Leslie	Carmichael	65032	Install second storage drive	Mon, Dec 13 2021	$175.00
			65036	Set up home network	Fri, Dec 17 2021	$220.22
			65044	Noisy fan	Wed, Dec 22 2021	$105.40
					Customer Total	$500.62
1010	Randall	Lemaire	65025	Troubleshoot hard drive noise	Mon, Dec 06 2021	$75.00
			65027	Replace hard drive with SSD	Tue, Dec 07 2021	$185.00
			65038	Install latest version of Windows	Mon, Dec 20 2021	$125.00
			65046	Windows 10 training	Tue, Dec 28 2021	$100.00
					Customer Total	$485.00

The report is grouped on the *CustID* field (displays with column heading *Customer ID*), allowing the owners to see how many work orders and how much revenue each customer generated.

Quick Steps

Group Records Using Report Wizard

1. Click Create tab.
2. Click Report Wizard button.
3. Choose table or query and fields.
4. Click Next.
5. If necessary, remove default grouped field name.
6. Double-click field name by which to group records.
7. Click Next.
8. Choose field(s) by which to sort.
9. Click Next.
10. Choose layout options.
11. Click Next.
12. Enter title for report.
13. Click Finish.

 Group & Sort

sum, average, maximum, minimum, or count for each group. For example, a report similar to the partial report shown in Figure 5.6, which organizes work orders by customer, allows the owners of RSR Computer Service to see which customer has provided the most revenue to their service business. In this report, the *CustID* field (column heading *Customer ID*) is used to group the records and a Sum function has been added to each group.

Recall from Level 1, Chapter 6 that either the Group & Sort button or the Report Wizard can be used to group records in a report. At the Report Wizard dialog box, shown in Figure 5.7, double-click a field name in the list box to add a grouping level. The preview window updates to display the grouped field in blue. More than one grouping level can be added to a report. To remove a grouping level, use the Remove Field button (the button with the left-pointing arrow) to remove the grouped level. Use the Priority buttons (the buttons with up and down arrows) to change the grouping order when there are multiple grouped fields.

If a report was not grouped when using the Report Wizard, group records can be added after the report has been generated using Layout view or Design view. In Layout view, click the Group & Sort button in the Grouping & Totals group on the Report Layout Tools Design tab. In Design view, click the Group & Sort button in the Grouping & Totals group on the Report Design Tools Design tab.

Figure 5.7 Grouping by a Field Using the Report Wizard

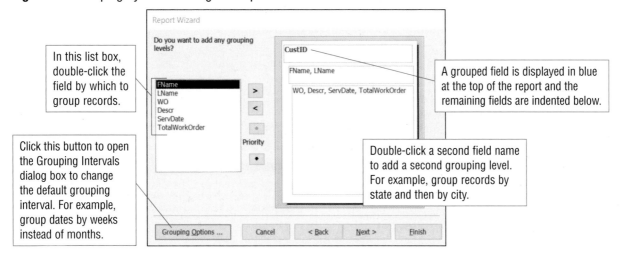

In this list box, double-click the field by which to group records.

Click this button to open the Grouping Intervals dialog box to change the default grouping interval. For example, group dates by weeks instead of months.

A grouped field is displayed in blue at the top of the report and the remaining fields are indented below.

Double-click a second field name to add a second grouping level. For example, group records by state and then by city.

Figure 5.8 Group, Sort, and Total Pane

Clicking the button in either view opens the Group, Sort, and Total pane, shown in Figure 5.8, at the bottom of the work area. Click the Add a group button and then click the field name by which to group records in the pop-up list.

Activity 3a Creating a Report with a Grouping Level Using the Report Wizard   Part 1 of 3

1. With **5-RSRCompServ** open, modify the TotalWorkOrders query to add two fields to include in a report by completing the following steps:

   a. Open the TotalWorkOrders query in Design view.

   b. Drag the *CustID* field from the *WorkOrders* table field list box to the *Field* row in the second column in the query design grid. *ServDate* and other fields will shift right to accommodate the new field.

   c. Drag the *Descr* field from the *WorkOrders* table field list box to the *Field* row in the third column in the query design grid.

   d. Save the revised query.

   e. Run the query.

   f. Close the query.

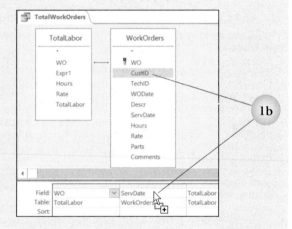

> In Step 1e, the revised query is run with the new fields added.

2. Create a report based on the TotalWorkOrders query that is grouped by service date by week using the Report Wizard by completing the following steps:

   a. Click the Create tab and then click the Report Wizard button in the Reports group.

   b. At the first Report Wizard dialog box with *Query: TotalWorkOrders* already selected in the *Tables/Queries* option box, move all the fields from the *Available Fields* list box to the *Selected Fields* list box and then click the Next button.

   c. At the second Report Wizard dialog box, specify that the report is to be grouped by the *ServDate* field by completing the following steps:

      1) With *CustID* displayed in blue in the preview section, indicating that the report will be grouped by customer number, click the Remove field button (the left-pointing arrow) to remove the grouping level.

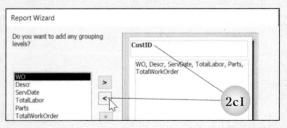

2) Double-click *ServDate* in the field
list box to add a grouping level by
the service date field. By default,
Access groups a date field by month.
3) Click the Grouping Options button.
4) At the Grouping Intervals dialog box
click the Grouping intervals option
box arrow, click *Week*, and then click
OK.
5) With the preview section now
displaying that the report will
be grouped by *ServDate by
Week*, click the Next button.

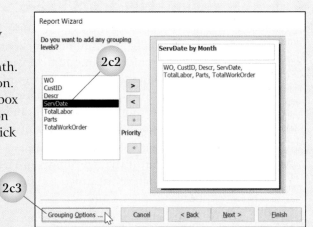

d. At the third Report Wizard dialog box,
click the arrow at the right of the first
sort option box, click *WO* at the drop-
down list to sort each group by work
order number in ascending order, and
then click the Next button.

e. At the fourth Report Wizard dialog
box, click *Landscape* in the *Orientation*
section and then click the Next button.
f. At the last Report Wizard dialog box, select the existing text in
the *What title do you want for your report?* text box, type WorkOrdersbyWeek, and then
click the Finish button.
3. Switch to Design view.
4. Add an alternate row color by completing the following steps:
a. Click the *Detail* section bar.
b. Click the Report Design Tools Format tab.
c. Click the Alternate Row Color button in the Background group.
d. Click *Blue-Gray, Accent 3, Lighter 80%* (seventh column, second row in the Theme Colors
section).
5. Switch to Report view.
6. Minimize the Navigation pane.
7. Preview the report and then switch to Layout view or Design view. Edit the text in the
report title and column heading labels, adjust column widths and the *Report Header* section
height as necessary until the report looks similar to the one shown below. The Organic
theme was applied in Activity 1d. Modify the colors for the *Report Header* and *Page Header*
sections to most closely match the theme colors shown below.

Work Orders by Week

Serv Date by Week	Work Order	CustID	Descr	Service Date	Total Labor	Parts	Total Work Order
49							
	65019	1025	Upgrade to new Windows	Wed, Dec 01 2021	$137.50	$0.00	$137.50
	65020	1030	Troubleshoot noisy fan	Thu, Dec 02 2021	$75.00	$72.50	$147.50
	65021	1035	Customer has blue screen upon boot	Thu, Dec 02 2021	$178.75	$0.00	$178.75
	65022	1040	Customer reports screen is fuzzy	Fri, Dec 03 2021	$82.50	$400.00	$482.50
	65023	1045	Upgrade RAM	Sat, Dec 04 2021	$62.50	$100.00	$162.50
50							
	65024	1005	Install malware protection	Mon, Dec 06 2021	$45.00	$125.00	$170.00
	65025	1010	Troubleshoot hard drive noise	Mon, Dec 06 2021	$75.00	$0.00	$75.00

8. Save the report.

Adding Functions to a Group

When a report is grouped, the Group, Sort, and Total pane can be used to add a calculation below a numeric field at the end of each group. Functions can also be added to more than one field within the group. For example, calculate a Sum function on a sales field and a Count function on an invoice field. The following functions are available for numeric fields: Sum, Average, Count Records, Count Values, Maximum, Minimum, Standard Deviation, and Variance. A non-numeric field can have a Count Records or Count Values function added.

The Group, Sort, and Total pane for a report with an existing grouping level looks similar to the one shown in Figure 5.9. Click the More Options button next to the group level to which a total is to be added to expand the available group options.

Click the option box arrow next to *with no totals* to open a *Totals* option box, similar to the one shown in Figure 5.10. Use the option boxes within this box to select the field a function should be added to and the type of aggregate function to calculate. Use the check boxes to choose to add a grand total to the end of the report, calculate group subtotals as percentages of the grand total, and decide whether to add the subtotal function to the *Group Header* or *Group Footer* section. Continue adding functions to other fields as needed and then click outside the *Totals* option box to close it.

Quick Steps

**Add Functions
to Group**
1. Open report in Design view or Layout view.
2. Click Group & Sort button.
3. Click More Options button.
4. Click arrow next to *with no totals.*
5. Choose field in *Total On* list box.
6. Choose function in *Type* list box.
7. If desired, click *Show Grand Total* check box.
8. If required, click *Show group subtotal as % of Grand Total* check box.
9. Click *Show subtotal in group header* or *Show subtotal in group footer* check box.
10. Repeat Steps 5–9 as needed for other fields.
11. Click outside *Totals* option box.
12. Close Group, Sort, and Total pane.

Figure 5.9 Group, Sort, and Total Pane with a Grouping Level Added

Click the More Options button to expand the pane to show group interval options, the *Totals* option, and other group options.

Figure 5.10 *Totals* Option Box in the Group, Sort, and Total Pane

Click here in the expanded options list to open the *Totals* option box, in which the field(s) and function(s) to add to the report are specified. Also add functions to calculate a grand total at the end of the report and group totals as percentages of the grand total.

1. With **5-RSRCompServ** open, display the WorkOrdersbyWeek report in Design view.
2. Add two functions at the end of each week to show the number of work orders and the total value of the work orders by completing the following steps:
   a. In the Grouping & Totals group on the Report Design Tools Design tab, click the Group & Sort button.

   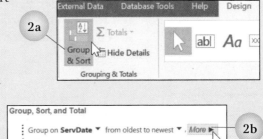

   b. At the Group, Sort, and Total pane at the bottom of the work area, click the More Options button next to *from oldest to newest* in the *Group on ServDate* group options.

   c. Click the option box arrow next to *with no totals* in the expanded group options.
   d. At the *Totals* option box with *WO* selected in the *Total On* option box, specify the type of function and the placement of the result by completing the following steps:
      1) Click the *Type* option box arrow and then click *Count Records* at the drop-down list.
      2) Click the *Show Grand Total* check box to insert a check mark. Access adds a Count function in a control object in the *Report Footer* section below the *WO* column.
      3) Click the *Show subtotal in group footer* check box to insert a check mark. Access displays a new section with the title *ServDate Footer* in the section bar below the *Detail* section and inserts a Count function in a control object below the *WO* column.

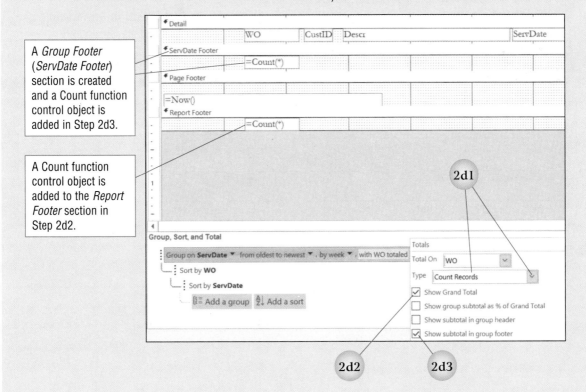

A *Group Footer* (*ServDate Footer*) section is created and a Count function control object is added in Step 2d3.

A Count function control object is added to the *Report Footer* section in Step 2d2.

e. With the *Totals* option box still open, click the *Total On* option box arrow and then click *TotalWorkOrder* at the drop-down list. The *Type* option defaults to *Sum* for a numeric field.

f. Click the *Show Grand Total* check box to insert a check mark. Access adds a Sum function in a control object in the *Report Footer* section.

g. Click the *Show subtotal in group footer* check box to insert a check mark. Access adds a Sum function in a control object in the *ServDate Footer* section.

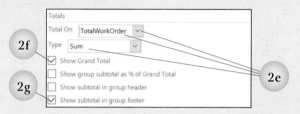

h. Click outside the *Totals* option box to close it.

3. Click the Group & Sort button to close the Group, Sort, and Total pane.

4. Review the two Count functions and two Sum functions added to the report in Design view.

5. Display the report in Print Preview to view the calculated results. Notice that the printout requires two pages and that the report's grand totals print on page 2. If any totals display with pound signs (#) switch to Design view, adjust the width of the control object, and then switch back to Print Preview. Also notice that Access added the Sum function below the Count function rather than at the bottom of the *Total Work Order* column.

6. Close Print Preview to switch back to Design view.

7. Click to select the Sum function control object in the *ServDate Footer* section, press and hold down the Shift key, click to select the Sum function control object in the *Report Footer* section, and then release the Shift key. Position the mouse pointer on the orange border of one of the selected control objects and then drag to move the two control objects simultaneously to the right below the TotalWorkOrder control object in the *Detail* section.

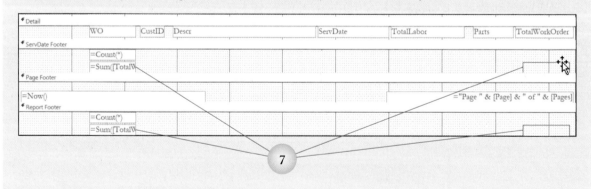

8. Add a label control object left of the Count function in the *ServDate Footer* section that displays the text *Work Order Count:* and another label control object left of the Sum function that displays the text *Weekly Total:*. Apply bold formatting and the red font color and right-align the text in the two label control objects. Resize and align the two label control objects as necessary. ***Note: Access displays an error flag on the two label control objects, indicating that these control objects are not associated with another control object. Ignore these error flags because the label control objects have been added for descriptive text only.***

9. Move up the Sum function control object and *Weekly Total:* label control object until they are at the same horizontal position as the Count function and then decrease the height of the *ServDate Footer* section as shown below.

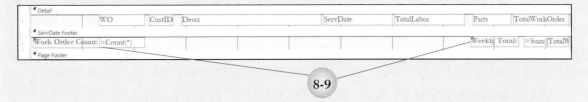

10. Display the report in Print Preview to view the labels. If necessary, return to Design view to further adjust the size and alignment.

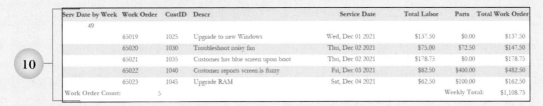

11. With the report displayed in Design view, select the Sum function control object in the *Report Footer* section and move up the control until it is positioned at the same horizontal position as the Count function below the *WO* column.

12. Click to select the *Work Order Count:* label control object, press and hold down the Shift key, click to select the *Weekly Total:* label control object, release the Shift key, click the Home tab, and then click the Copy button. Click the *Report Footer* section bar and then click the Paste button. Move and align the copied labels as shown below. Edit the *Weekly Total:* label control object to *Grand Total:* as shown.

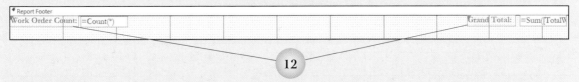

13. Display the report in Report view. Scroll to the bottom of the page to view the labels next to the grand totals. If necessary, return to Design view to further adjust the size and alignment and then save and close the report.

Check Your Work

Adding a Calculated Field to a Report

Several options are available for adding a calculated field to a report. Either create a query that includes a calculated column and then create a new report based on the query or create a calculated control object in an existing report using Design view. A calculated value can be placed in any section of the report. To create a calculated control object in a report, follow the same steps learned in Chapter 4 for adding a calculated control object to a form. Remember that field names in a formula are enclosed in square brackets.

A calculated control object can be used as a field in another formula. To do this, reference the calculated control object in the formula by its Name property (found on the Other tab of the Property Sheet task pane). If necessary, change the name of the calculated control object to a more descriptive or succinct name.

Activity 3c Adding a Calculated Field to a Report   Part 3 of 3

1. With **5-RSRCompServ** open, display the WorkOrdersbyWeek report in Design view. In the following steps you will create a calculated control object to project the value of January's work orders. Assume that the value for January will be a 10% increase over the value for December noted in the grand total.
2. Change the name of the text box control object that contains the summed value of December's work orders to a shortened name by completing the following steps:
   a. Click the control object that contains the summed monthly value (displays part of the formula =*Sum([TotalWorkOrder])*) in the *Report Footer* section.
   b. Click the Property Sheet button in the Tools group on the Report Design Tools Design tab.
   c. Click the Other tab in the Property Sheet task pane.
   d. Select and delete the existing text (displays *AccessTotalsTotal Work Order*) in the *Name* property box, type MonthlyTotal, and then close the Property Sheet task pane.
3. Add a calculated text box control object to the *Report Footer* section by completing the following steps:
   a. Click the Text Box button in the Controls group.
   b. Position the crosshairs with the text box icon attached at the 6-inch mark on the horizontal ruler in the *Report Footer* section. Drag to create a control object that is approximately 2 inches wide by 0.25 inch tall at the same vertical position as the other control objects in the *Report Footer* section and then release the mouse.
   c. Click in the text box control object (displays *Unbound*), type =[MonthlyTotal]*1.1, and then press the Enter key.

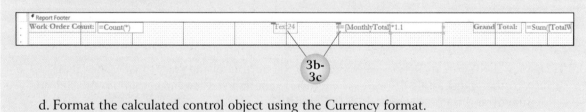

   d. Format the calculated control object using the Currency format.

e. Select the entry in the label control object (displays *Textxx*, where *xx* is the text box control object number) and then type January's Projected Total:. The two control objects will be resized and moved in steps below.

f. Apply bold formatting and the red font color to the text in both control objects.

g. Resize the text box control that contains the new formula until the left edge of the control object is aligned at approximately the 7-inch position on the horizontal ruler.

h. Resize the label control object that contains the text *January's Projected Total:* until the right edge of the control object meets the left edge of the text box control object, as shown below. Right-align the text in the label control object.

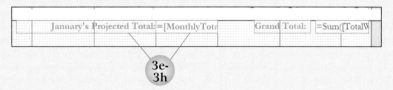

4. Display the report in Report view. Scroll to the bottom of the page to view the calculated field. If necessary, return to Design view to further adjust the size and alignment including decreasing the height of the *Report Footer* section and then save and close the report. Further changes to the report will be made in Activity 4.

5. Redisplay the Navigation pane.

 Check Your Work

---

Activity 4 **Modify Section and Group Properties**   **1 Part**

You will change a report's page setup and then modify section and group properties to control print options.

 Tutorial

Modfying Section
Properties

Quick Steps

**Modify Section
Properties**
1. Open report in
   Design view.
2. Double-click white
   section bar.
3. Change properties.
4. Close Property Sheet
   task pane.

Modifying Section Properties

A report has a Property Sheet, each control object within a report has a Property Sheet, and each section within a report has a Property Sheet. Section properties control whether the section is visible when printed, along with the section's height, background color, special effects, and so on.

Figure 5.11 displays the Format tab in the Property Sheet task pane for the *Report Header* section. Some of the options can be changed without opening the Property Sheet task pane. For example, increase or decrease the height of a section by dragging the top or bottom of a white section bar in Design view. The background color can also be set using the Fill/Back Color button in the Font group on the Report Design Tools Format tab.

Figure 5.11 Report Header Section Property Sheet Task Pane with the Format Tab Selected

Use the Keep Together property to ensure that a section does not split over two pages because of a page break. If necessary, Access prints the section at the top of the next page. However, if the section is longer than can fit on one page, Access continues printing the section on the following page. In that case, decrease the margins and/or apply a smaller font size to fit all the text for the section on one page.

Use the Force New Page property to insert a page break before a section begins *(Before Section)*, after a section ends *(After Section)*, or before and after a section *(Before & After)*.

Tutorial

Keeping a Group Together on the Same Page

Quick Steps

Keep Group Together on One Page
1. Open report in Design view or Layout view.
2. Click Group & Sort button.
3. Click More Options button.
4. Click arrow next to *do not keep group together on one page*.
5. Click print option.
6. Close Group, Sort, and Total pane.

Keeping a Group Together on the Same Page

Open the Group, Sort, and Total pane and click the More Options button for a group to specify whether to keep a group together on the same page. By default, Access does not keep a group together. Click the option box arrow next to *do not keep group together on one page* and then click a print option, as shown in Figure 5.12.

Figure 5.12 Group, Sort, and Total Pane with Keep Group Together Print Options

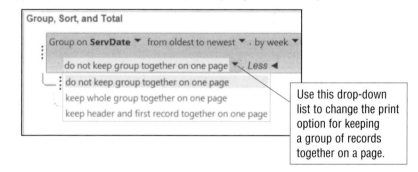

Use this drop-down list to change the print option for keeping a group of records together on a page.

1. With **5-RSRCompServ** open, display the WorkOrdersbyWeek report in Print Preview and then click the Zoom button (not the button arrow) in the Zoom group to change the zoom to view an entire page within the window.
2. Click the Next Page button in the Page Navigation bar to view page 2 of the report, which shows the grand total. Notice that week 52 is split over two pages.
3. Switch to Design view and then minimize the Navigation pane.
4. Change the section properties for the *ServDate Footer* and *Report Footer* sections displaying the Count and Sum functions and control objects by completing the following steps:
   a. Double-click the *ServDate Footer* section bar to open the section's Property Sheet task pane.
   b. Click in the *Back Color* property box on the Format tab and then click the Build button to open the color palette.
   c. Click *Blue-Gray, Accent 3, Lighter 80%* (seventh column, second row in the *Theme Colors* section).
   d. Close the Property Sheet task pane.
   e. Select the Count function control object and Sum function control object in the *ServDate Footer* section and then change the font color to red and apply bold formatting.

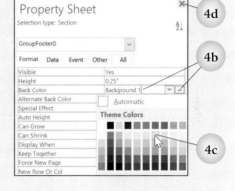

   f. With the Count and Sum function control objects still selected, right-click one of the selected control objects, point to *Fill/Back Color* at the shortcut menu, and then click *Transparent*. The background color applied in Step 4c will now display behind the calculations.
   g. Double-click the *Report Footer* section bar, change the Back Color property to the same color as the *ServDate Footer* section (see Steps 4b and 4c), and then close the Property Sheet task pane.
   h. Apply the formatting in Steps 4e and 4f to the Count and Sum function control objects in the *Report Footer* section.
   i. Right-click the text box control object containing the formula *=[MonthlyTotal]*1.1* and then click *Properties* at the shortcut menu. Change the Back Style property to *Transparent* and the Border Style property to *Transparent*. Close the Property Sheet task pane.
5. Print the work orders on a separate page by completing the following steps:
   a. Click the Group & Sort button on the Report Design Tools Design tab.
   b. Click the More Options button in the Group, Sort, and Total pane.
   c. Click the option box arrow next to *do not keep group together on one page* and then click *keep whole group together on one page* at the drop-down list.

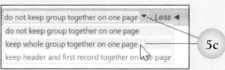

   d. Close the Group, Sort, and Total pane.
6. Create a label control object at the top right of the *Report Header* section with your first and last names. Apply bold formatting to the label control object and click the *Blue-Gray, Accent 3, Darker 50%* font color (seventh column, last row in the *Theme Colors* section).
7. Display the report in Print Preview and then change the zoom to view an entire page within the window. Compare the report with the one shown in Figure 5.13. Scroll to page 2 to view all of work orders for week 52 on the same page.
8. Save, print, and then close the report.
9. Redisplay the Navigation pane.

Check Your Work

Figure 5.13 Page 1 of the Completed Report in Activity 4

Work Orders by Week							Student Name
Serv Date by Week	Work Order	CustID	Descr	Service Date	Total Labor	Parts	Total Work Order
49							
	65019	1025	Upgrade to new Windows	Wed, Dec 01 2021	$137.50	$0.00	$137.50
	65020	1030	Troubleshoot noisy fan	Thu, Dec 02 2021	$75.00	$72.50	$147.50
	65021	1035	Customer has blue screen upon boot	Thu, Dec 02 2021	$178.75	$0.00	$178.75
	65022	1040	Customer reports screen is fuzzy	Fri, Dec 03 2021	$82.50	$400.00	$482.50
	65023	1045	Upgrade RAM	Sat, Dec 04 2021	$62.50	$100.00	$162.50
Work Order Count:	5					Weekly Total:	$1,108.75
50							
	65024	1005	Install malware protection	Mon, Dec 06 2021	$45.00	$125.00	$170.00
	65025	1010	Troubleshoot hard drive noise	Mon, Dec 06 2021	$75.00	$0.00	$75.00
	65026	1025	Upgrade RAM	Tue, Dec 07 2021	$37.50	$100.00	$137.50
	65027	1010	Replace hard drive with SSD	Tue, Dec 07 2021	$75.00	$110.00	$185.00
	65028	1030	Reinstall operating system	Wed, Dec 08 2021	$112.50	$0.00	$112.50
	65029	1035	Set up automatic backup	Fri, Dec 10 2021	$22.50	$0.00	$22.50
	65031	1045	Customer reports noisy hard drive	Sat, Dec 11 2021	$87.50	$0.00	$87.50
Work Order Count:	7					Weekly Total:	$790.00
51							
	65030	1000	Clean malware from system	Mon, Dec 13 2021	$40.00	$0.00	$40.00
	65032	1008	Install second storage drive	Mon, Dec 13 2021	$100.00	$75.00	$175.00
	65033	1000	Install Windows 10	Tue, Dec 14 2021	$162.50	$335.75	$498.25
	65034	1035	File management training	Tue, Dec 14 2021	$75.00	$0.00	$75.00
	65035	1020	Office 365 training	Thu, Dec 16 2021	$125.00	$0.00	$125.00
	65036	1008	Set up home network	Fri, Dec 17 2021	$135.00	$85.22	$220.22
	65037	1015	Biannual computer maintenance	Fri, Dec 17 2021	$62.50	$8.75	$71.25
	65039	1030	Set up automatic backup	Sat, Dec 18 2021	$30.00	$0.00	$30.00
Work Order Count:	8					Weekly Total:	$1,234.72

Tuesday, January 19, 2021   Page 1 of 2

Activity 5 Create and Format a Chart   1 Part

You will create a chart in a customer report to show the total parts and labor on work orders by customer for a month.

Inserting, Editing, and Formatting a Chart in a Report

A chart can be added to a report to graphically display the numerical data from another table or query. The chart is linked to a field in the report that is common to both objects. Access summarizes and graphs the data from the charted table or query based on the fields selected for each record in the report.

Inserting a Chart

Insert Chart

Inserting a Chart

With a report open in Design view, increase the height or width of the *Detail* section to make room for the chart, click the Insert Chart button in the Controls group on the Report Design Tools Design tab, and then drag the crosshairs with the chart icon attached to create the approximate height and width of the chart. Release the mouse and Access launches the Chart Wizard, which has six dialog boxes that guide the user through the steps of creating a chart. The first Chart Wizard dialog box is shown in Figure 5.14.

1. Open report in Design view.
2. Click Insert Chart button.
3. Drag to create control object of required height and width.
4. Select table or query for chart data.
5. Click Next.
6. Add fields to use in chart.
7. Click Next.
8. Click chart type.
9. Click Next.
10. Add fields as needed to chart layout.
11. Click Preview Chart.
12. Close Sample Preview window.
13. Click Next.
14. Select field to link report with chart.
15. Click Next.
16. Type chart name.
17. Click Finish.

Figure 5.14 First Chart Wizard Dialog Box

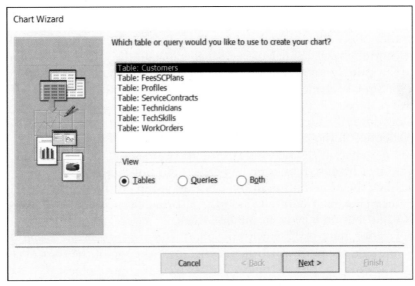

Activity 5 uses the Chart Wizard to insert a chart in a customer report that depicts the total value of work orders for each customer by month. The data for the chart will be drawn from a related query. A chart can also be inserted into a form and formatted by completing steps similar to those in Activity 5.

Activity 5 Creating a Report and Inserting a Chart   Part 1 of 1

1. With **5-RSRCompServ** open, create a new report using the Report Wizard by completing the following steps:
   a. Click *Customers* in the Tables group in the Navigation pane, click the Create tab, and then click the Report Wizard button.
   b. At the first Report Wizard dialog box with *Table: Customers* selected in the *Tables/Queries* option box, move the *CustID*, *FName*, *LName*, and *ServCont* fields from the *Available Fields* list box to the *Selected Fields* list box and then click the Next button.
   c. At the second Report Wizard dialog box, with no group field selected, click the Next button.
   d. Click the Next button at the third Report Wizard dialog to choose not to sort the report.
   e. Click *Columnar* at the fourth Report Wizard dialog box and then click the Next button.
   f. Click at the end of the current text in the *What title do you want for your report?* text box, type WOChart so that the report title is *CustomersWOChart*, and then click the Finish button.
2. Minimize the Navigation pane and display the report in Design view.
3. To minimize the number of pages printed, change the page layout of the report to two columns by completing the following steps:
   a. Delete the date and page number control objects in the *Page Footer* section.
   b. Drag the right side of the grid to the left to meet the right edge of the *LName* text box control object. The grid will be approximately 3.8 inches wide.
   c. Click the Columns button in the Page Layout group on the Report Design Tools Page Setup tab, change *1* to *2* in the *Number of Columns* text box, and then click OK.
   d. Switch to Print Preview and review the report.

4. There is not enough room to place a chart showing the value of the work orders for each customer. Switch the report back to one column by completing the following steps:
   a. Close Print Preview and press the keys Ctrl + Z three times or until the two text box control objects are deleted and the change in the grid size is undone.
   b. Click the Columns button in the Page Layout group on the Report Design Tools Page Setup tab, change *2* to *1* in the *Number of Columns* text box, and then click OK.
5. Select the text in the report title and then type December 2021 Work Orders by Customer.
6. Drag down the top of the *Page Footer* section bar until the bottom of the *Detail* section is positioned at the 2-inch mark on the vertical ruler.
7. Insert a chart at the right side of the report to show the value of the work orders for each customer by month by completing the following steps:
   a. Click the More button in the Controls group on the Report Design Tools Design tab and then click the Chart button in the last row. (Depending on the size of your screen, the Chart button may be in the first row).
   b. Position the crosshairs with the chart icon attached in the *Detail* section at the 5-inch mark on the horizontal ruler aligned near the top of the *CustID* control object and then drag down and to the right to create a chart control object of the approximate height and width shown below.

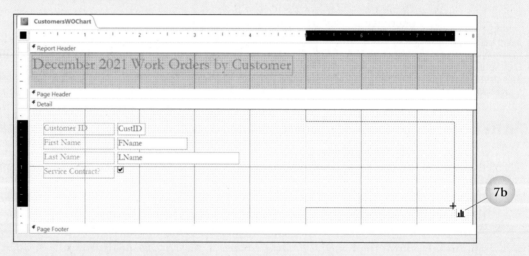

   c. At the first Chart Wizard dialog box, click *Queries* in the *View* section, click *Query: TotalWorkOrders* in the list box, and then click the Next button.

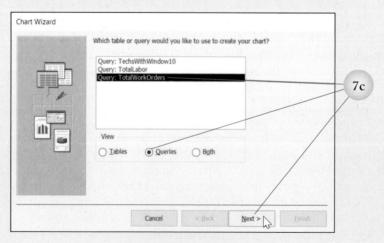

d. At the second Chart Wizard dialog box, double-click *ServDate* and *TotalWorkOrder* in the *Available Fields* list box to move the fields to the *Fields for Chart* list box and then click the Next button.

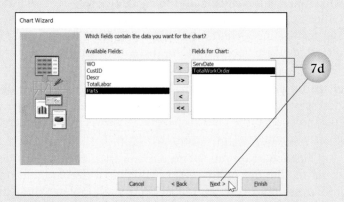

e. At the third Chart Wizard dialog box, click the second chart type in the first row (*3-D Column Chart*) and then click the Next button.

f. At the fourth Chart Wizard dialog box, look at the fields that Access has already placed to lay out the chart. Since only two fields were added, Access automatically used the numeric field with a Sum function as the data series for the chart and the date field as the *x*-axis category. Click the Next button.

g. At the fifth Chart Wizard dialog box, notice that Access has correctly detected *CustID* as the field to link records to in the Customers report with the chart (based on the TotalWorkOrders query). Click the Next button.

h. At the last Chart Wizard dialog box, click the *No, don't display a legend option* and then click the Finish button. Access inserts a chart within the height and width of the chart control. The chart displayed in the control object in Design view is not the actual chart based on the query data; it is only a sample to show the chart elements.

8. Display the report in Print Preview and scroll through the four pages. Customers for which an empty chart displays have no work order data to be graphed.

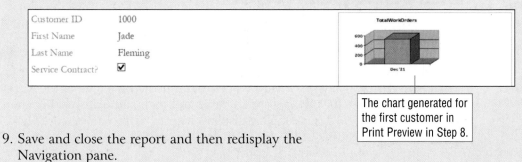

The chart generated for the first customer in Print Preview in Step 8.

9. Save and close the report and then redisplay the Navigation pane.

Editing and Formatting a Chart

The Chart feature within Access is not the same Chart tool available in Word, Excel, and PowerPoint. Access uses the Microsoft Graph application for charts, which means the editing and formatting processes are different than in other Office applications. Depending on your version of Access, use care when editing and formatting a chart, as changes can result in the chart dimensions becoming distorted. If this occurs, delete the control object and start again.

Open a report in Design view and then double-click a chart object to edit the chart. In chart-editing mode, a Menu bar and a toolbar display at the top of the Access window, as well as a datasheet for the chart in the work area. Use these tools to change the chart type; add, remove, or change chart options; and format chart elements.

Click Chart on the Menu bar and then click *Chart Options* at the drop-down menu to add, delete, or edit text in chart titles and add or remove chart axes, gridlines, the legend, data labels, or a data table at the Chart Options dialog box. Click *Chart Type* at the Chart drop-down menu to open the Chart Type dialog box and choose a different type of chart, such as a bar chart or pie chart.

Right-click an object within a chart—such as the chart title, legend, chart area, or data series—and a format option displays in the shortcut menu for the selected chart element. Click the format option to open a Format dialog box for the selected element. Make the required changes and then click OK.

After editing the chart, click outside the chart object to exit chart-editing mode. Sometimes, Access displays a sample chart within the control object in chart-editing mode instead of the actual chart, which can make editing specific chart elements difficult if the chart does not match the sample. If this occurs, exit chart-editing mode, close and reopen the report in Design view, or switch views to cause Access to update the chart displayed in the control object.

Quick Steps

Change Chart Options
1. Open report in Design view.
2. Double-click chart.
3. Click Chart on Menu bar.
4. Click *Chart Options*.
5. Click tab.
6. Change options.
7. Click OK.

Change Chart Type
1. Open report in Design view.
2. Double-click chart.
3. Click Chart on Menu bar.
4. Click *Chart Type*.
5. Click chart type in list box.
6. Click chart subtype.
7. Click OK.

Format Chart Element
1. Open report in Design view.
2. Double-click chart.
3. Right-click chart element.
4. Click *Format*.
5. Change format options.
6. Click OK.

Activity 6 Create a Blank Report with Hyperlinks, and a List Box   1 Part

You will use the Blank Report tool to create a new report for viewing technician certifications. In the report, you will reorder the tab fields, create a list box inside a tab control object, change the shape of the tab control object, and add hyperlinks.

 Tutorial

Creating a Report Using the Blank Report Tool

 Blank Report

Quick Steps

Create Blank Report
1. Click Create tab.
2. Click Blank Report button.
3. Expand field list for table.
4. Drag fields to report.
5. Save report.

Creating a Report Using the Blank Report Tool

A blank report created with the Blank Report tool begins with no control objects or formatting and displays as a blank white page in Layout view, similar to a form created with the Blank Form tool, as learned in Chapter 4. Click the Create tab and then click the Blank Report button in the Reports group to begin a new report. Access opens the Field List task pane at the right side of the work area. Expand the list for the appropriate table and then add fields to the report as needed. If the Field List task pane displays with no table names, click the <u>Show all tables</u> hyperlink at the top of the task pane.

Adding a Tab Control Object to a Report

 Tab Control

Recall from Chapter 4 that a tab control object can be added to a form to display fields from different tables on separate pages. In Form view, a page is displayed by clicking the page tab. Similarly, a tab control object can be used in a report to display fields from the same table or a different table in pages. To create a tab control object in a report, follow the steps from Chapter 4 for adding a tab control object to a form.

Adding a List Box or Combo Box to a Report

Tutorial

Adding a List Box to a Report

 List Box

Combo Box

Like a list box in a form, a list box in a report displays a list of values for a field within the control object. In Report view, the entire list for the field can easily be seen. If a list is too long for the size of the list box control object, scroll bars display that can be used to scroll up or down the list when viewing the report. Although the data cannot be edited in a report, a list box can be used to view all the field values and see which values have been selected for the current record.

A combo box added to a report does not display as a list. However, the field value that was entered into the field from the associated table, query, or form is shown in the combo box. Since the data cannot be edited in a report, the combo box field is not shown as a drop-down list. A combo box can be changed to display as a list box within the report. In this case, the list box displays all the field values and the value stored in the current record shown is selected within the list. A list box or combo box can be added to a report by following the steps from Chapter 4 for adding a list box or combo box to a form.

Adding Hyperlinks to a Report

Use the Hyperlink button in the Controls group on the Report Layout Tools Design tab to create a link in a report to a web page, graphic, email address, or program. Click the Hyperlink button and then click within the report to open the Insert Hyperlink dialog box, in which the user provides the text to display in the control object and the address to which the object should be linked. Use the Places bar left of the Insert Hyperlink dialog box to choose to link to an existing file or web page, another object within the database, or an email address.

Changing the Shape of a Control Object

The Change Shape button in the Control Formatting group on the Report Layout Tools Format tab contains a drop-down list with eight shape options. Use the shape options to modify the appearance of a command button, toggle button, navigation button, or tab control object. Select the control object to be modified, click the Change Shape button, and then click a shape at the drop-down list.

Changing the Tab Order of Fields

 Hyperlink

 Change Shape

 Tab Order

Recall from Chapter 4 how to open the Tab Order dialog box and change the order in which the Tab key moves from field to field. In a report, the order in which the Tab key moves from field to field in Report view can also be changed. Although data is not added, deleted, or edited in Report view, use the Tab key to move within a report. Display the report in Design view, click the Tab Order button in the Tools group on the Report Design Tools Design tab, and then drag the fields up or down the *Custom Order* list.

1. With **5-RSRCompServ** open, click the Create tab and then click the Blank Report button in the Reports group. Save the report with the name *TechCertifications*.
2. If the Field List task pane at the right side of the work area does not display the table names, click the <u>Show all tables</u> hyperlink. If the table names display, proceed to Step 3.
3. Add fields from the Technicians table and a tab control object to the report by completing the following steps:
   a. Click the plus symbol next to the *Technicians* table name to expand the field list.
   b. Click *TechID* in the Field List task pane and then drag the field to the top left of the report.
   c. Right-click in the *Technician ID* column, point to *Layout* at the shortcut menu, and then click *Stacked*. A stacked layout is better suited to this report because it shows each technician's certifications next to his or her name in a tab control object.
   d. Drag the *FName* field from the Field List task pane below the first *Technician ID* text box control object. Release the mouse when the pink bar is below the *01* text box control object next to *Technician ID*.

   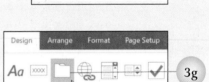

   e. Drag the *LName* field from the Field List task pane below the first *First Name* text box control object. Release the mouse when the pink bar displays below *Pat*.
   f. Select the *HPhone* and *CPhone* fields in the Field List task pane and then drag the two fields below the *Last Name* text box control object. Release the mouse when the pink bar displays below *Hynes*.

   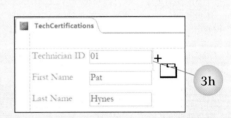

   g. Click the Tab Control button in the Controls group on the Report Layout Tools Design tab.
   h. Position the mouse pointer with the tab control icon attached right of the first *Technician ID* text box control object in the report. Click the mouse when the pink bar displays right of the *01* text box control object.
   i. Right-click the selected tab control object, point to *Layout*, and then click *Remove Layout* at the shortcut menu.
   j. With the first *Pagexx* tab selected (where *xx* is the page number) point to the bottom orange border of the selected tab control object until the pointer displays as an up-and-down-pointing arrow and then drag down the bottom of the object until it aligns with the bottom of the *Cell Phone* control object.

   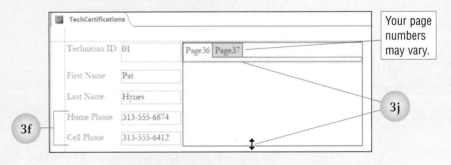

4. Add a field from the TechSkills table to the tab control object and change the control to a list box by completing the following steps:
   a. Click the plus symbol next to the *TechSkills* table name in the *Fields available in related tables* section of the Field List task pane to expand the list.
   b. Click to select the *Certifications* field name and then drag the field to the first page in the tab control object, next to *Technician ID 01*.
   c. Access inserts the field in the page with both the label control object and text box control object selected. To change the field to display as a list box, click to select only the text box control object (displays *CCNA Cloud, CCNA Wireless,* and *Microsoft MCT* in the first record).
   d. Right-click the selected text box control object, point to *Change To*, and then click *List Box* at the shortcut menu.

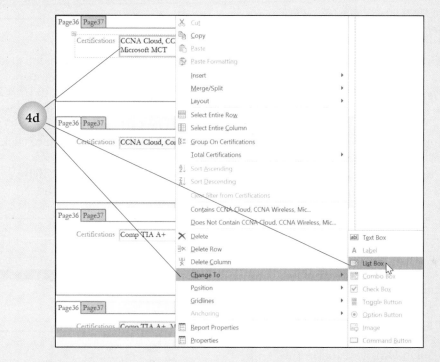

5. Add a hyperlink to the bottom of the tab control object by completing the following steps:
   a. Click the Hyperlink button in the Controls group on the Report Layout Tools Design tab.
   b. Position the mouse pointer with the hyperlink icon attached below *Certifications* in the tab control object. Click the mouse when the pink bar displays.
   c. At the Insert Hyperlink dialog box, click in the *Text to display* text box and then type Cisco Certifications.
   d. Click in the *Address* text box, type https://www.cisco.com/c/en/us/training-events/training-certifications/overview.html, and then click OK.
   e. Drag the right orange border of the hyperlink control object to the right until all the text displays within the control object.

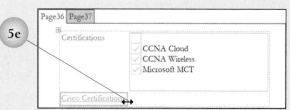

6. Click the second page in the tab control to select it, right-click, and then click *Delete Page* at the shortcut menu. Do not be concerned if Access displays the first page with an empty list box. The screen will refresh at the next step.

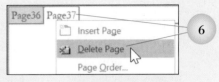

7. Double-click over *Pagexx* (where *xx* is the page number) in the tab control object to select the entire tab control object. Click the Report Layout Tools Format tab, click the Change Shape button in the Control Formatting group, and then click *Rectangle: Single Corner Snipped* at the drop-down list (third column, second row).

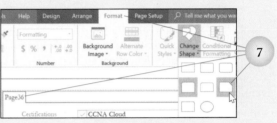

8. Click to select *Pagexx* (where *xx* is the page number), click the Report Layout Tools Design tab, and then click the Property Sheet button to open the Property Sheet task pane. With *Selection type: Page* displayed at the top of the Property Sheet task pane, click in the *Caption* property box, type Technician's Certifications, and then close the Property Sheet task pane.

9. Right-click the *Technician ID* text box control object (displays *01*) and then click *Select Entire Column* at the shortcut menu. Click the Report Layout Tools Format tab, click the Shape Outline button, and then click *Transparent* at the drop-down list.

10. Switch to Report view. Click in the first *Technician ID* text box and press the Tab key four times to see how active field moves through the report in order starting at the left side of the report.

11. Switch to Design view. You want the Tab key to move to the technician's certifications after their cell phone. The *TabCtrlxx* field (where *xx* is the control number) will be moved to the last field in the order. Change the tab order of the fields by completing the following steps:

    a. Click the Tab Order button in the Tools group on the Report Design Tools Design tab.

    b. At the Tab Order dialog box, click in the bar next to *TabCtrlxx* field (where *xx* is the control number) in the *Custom Order* section to select the *TabCtrlxx* field and then drag the field to the bottom of the list until the black line displays below *CPhone*.

    c. Click OK.

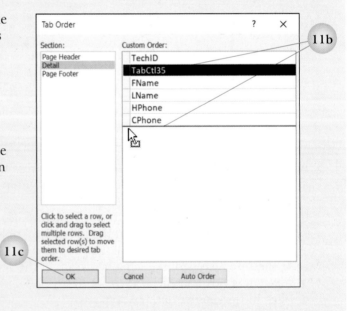

12. Save the revised report and then switch to Report view. Press the Tab key. Notice that the first field selected is the *Technician ID* field. Press the Tab key a second time. Notice that the selected field moves to *First Name*. Press the Tab key six more times to watch the selected fields move through the fields.

13. Click the Cisco Certifications hyperlink to open a Microsoft Edge window and display the Cisco Training & Certifications web page.

14. Close the browser window.
15. Display the report in Print Preview, print only the first page, and then close Print Preview.
16. Save and close the TechCertifications report and then close **5-RSRCompServ**.

Chapter Summary

- Click the Create tab and then click the Report Design button to create a custom report using Design view.

- A report can contain up to five sections: *Report Header, Page Header, Detail, Page Footer,* and *Report Footer*.

- *Group Header* and *Group Footer* sections can be added to group records that contain repeating values in a field, such as a department or city.

- Connect a table or query to a report using the *Record Source* property box in the Data tab of the report's Property Sheet task pane.

- Display the Field List task pane and then drag individual fields or a group of fields from the table or query to the *Detail* section of the report, which represents the body of the report.

- Place label control objects to be used as column headings in a tabular report within the *Page Header* section.

- Apply a theme to a report to maintain a consistent look to all reports.

- A related table or query can be inserted as a subreport within a main report. A subreport is stored as a separate object outside the main report.

- Click the Page Numbers button in the Header / Footer group to open the Page Numbers dialog box, where the format, position, and alignment of page numbers in a report are specified.

- The current date and/or time can be added as a control object within the *Report Header* section using the Date and Time button in the Header / Footer group.

- Add online graphics and draw lines in a report using the same techniques for adding graphics and drawing lines in a form.

- A report can be grouped by a field at the Report Wizard or by opening the Group, Sort, and Total pane.

- Functions such as Sum, Average, and Count can be added to each group within a report and grand totals can be added to the end of a report by expanding the group options in the Group, Sort, and Total pane.

- Create a calculated control object in a report using the Text Box button in the Controls group.

- Each section within a report has a set of properties that can be viewed or changed by opening the section's Property Sheet task pane.

- Use the Keep Together property to prevent a section from splitting across two pages.

- Use the Force New Page property to automatically insert a page break before a section begins, after a section ends, or before and after a section.

- At the Group, Sort, and Total pane, specify whether to keep an entire group together on the same page.

- A chart can be added to a report to graphically display numerical data from another table or query.

- A report can be formatted into multiple columns using the Columns button in the Report Design Tools Page Setup tab.

- To create a chart in a report, open the report in Design view, click the Chart button in the Controls group, and then use the Chart Wizard to generate the chart.

- Open a report in Design view and then double-click a chart control object to edit the chart using Microsoft Graph by changing the chart type; adding, removing, or changing chart options; and formatting chart elements.

- Use the Blank Report tool in the Reports group on the Create tab to create a new report with no control objects or formatting applied. The report opens as a blank white page in Layout view; the Field List task pane opens right of the work area.

- A tab control object, list box, and combo box can be added to a blank report using the same techniques for adding these control objects to a form.

- Use the Hyperlink button to create a link in a report to a web page, graphic, email address, or program.

- Modify the shape of a command button, toggle button, navigation button, or tab control object using the Change Shape button in the Control Formatting group on the Report Layout Tools Format tab.

- Change the order in which the Tab key moves from field to field within a report at the Tab Order dialog box. Display a report in Design view and click the Tab Order button in the Tools group on the Report Design Tools Design tab.

- As you become more comfortable with reports, explore other tools available in Layout View and Design view using the Design, Arrange, Format, and Page Setup tabs. More features are available to assist you with creating professional-quality reports.

Commands Review

FEATURE	RIBBON TAB, GROUP	BUTTON	KEYBOARD SHORTCUT
add existing fields	Report Design Tools Design, Tools		
blank report	Create, Reports		
change shape of selected control	Report Layout Tools Format, Control Formatting		
date and time	Report Design Tools Design, Header / Footer		
Design view	Report Design Tools Design, Views		
group and sort	Report Design Tools Design, Grouping & Totals		
insert chart	Report Design Tools Design, Controls		
insert hyperlink	Report Layout Tools Design, Controls		
insert image	Report Design Tools Design, Controls		
page numbers	Report Design Tools Design, Header / Footer		
Property Sheet task pane	Report Design Tools Design, Tools		F4
report design	Create, Reports		
Report view	Report Design Tools Design, Views		
Report Wizard	Create, Reports		
subreport	Report Design Tools Design, Controls		
tab control object	Report Layout Tools Design, Controls		
tab order	Report Design Tools Design, Tools		
theme	Report Design Tools Design, Themes		
title control object	Report Design Tools Design, Header / Footer		

Microsoft®

Access®

Using Access Tools and Managing Objects

Performance Objectives

Upon successful completion of Chapter 6, you will be able to:

1 Create a new database using a template

2 Save a database as a template

3 Add prebuilt objects to a database using an Application Parts template

4 Create a new form using an Application Parts Blank Form

5 Create a form to be used as a template in a database

6 Create a table by copying the structure of another table

7 Evaluate a table using the Table Analyzer Wizard

8 Evaluate a database using the Performance Analyzer

9 Split a database

10 Print documentation about a database using the Database Documenter

11 Rename and delete objects within a database file

12 Use SQL to modify a query

Access provides tools to assist with creating and managing databases. Use a template to create a new database or a new table and/or a related group of objects. A blank form template provides a predefined layout and may include a form title and command buttons. If none of the predefined templates is suitable, create a custom template. Access provides wizards to assist with analyzing tables and databases to improve performance. A database can be split into two files to store the tables separately from the queries, forms, and reports. Use the Database Documenter to print a report that provides details about objects and their properties. In this chapter, you will learn how to use these Access tools and how to rename and delete objects in the Navigation pane.

 Data Files

Before beginning chapter work, copy the AL2C6 folder to your storage medium and then make AL2C6 the active folder.

The online course includes additional training and assessment resources.

You will create a new database using one of the database templates supplied with Access and create your own template.

 Tutorial

Creating a New Database Using a Template

Quick Steps

Create Database from Template

1. Start Access.
2. Click template.
3. Click Browse button.
4. Navigate to drive and/or folder.
5. Edit file name as required.
6. Click OK.
7. Click Create button.

Creating a New Database Using a Template

At the Recent or New backstage area, create a new database using one of the professionally designed templates provided by Microsoft. The database templates provide a complete series of objects, including predefined tables, forms, reports, queries, and relationships. Use a template as provided and immediately start entering data or base a new database on a template and modify the objects as needed. If a template exists for a database application that is needed, save time by basing the database on one of the template designs.

To create a new database using a template, start Access and then click one of the available templates like the ones shown in Figure 6.1 in the Recent backstage area. If Access is already open, click one of the available templates at the New backstage area. The templates are constantly being updated, so the options you see may vary. When the template is clicked, a preview appears, similar to what is shown in Figure 6.2. The name of the template provider, a description of the template, and the download size are displayed in the preview. Use the directional arrows on either side of the preview to move through the other available templates. Once a template has been chosen, click the Browse button to navigate to the drive and/or folder where the database is to be stored and then type a file name at the *File Name* text box. Click the Create button to create the database.

If none of the sample templates is suitable for the type of database to be created, close the preview and search for templates online by clicking one of the hyperlinked suggested searches (such as <u>Database</u>, <u>Business</u>, or <u>Logs</u>). Another option is to type search words in the search text box. Templates are downloaded from Office.com.

Figure 6.1 Available Templates

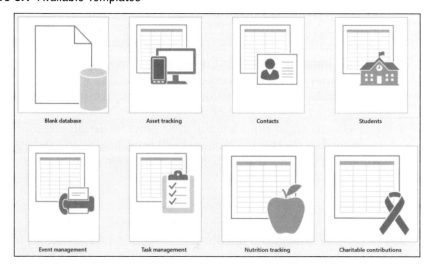

Figure 6.2 Available Templates in the New Backstage Area

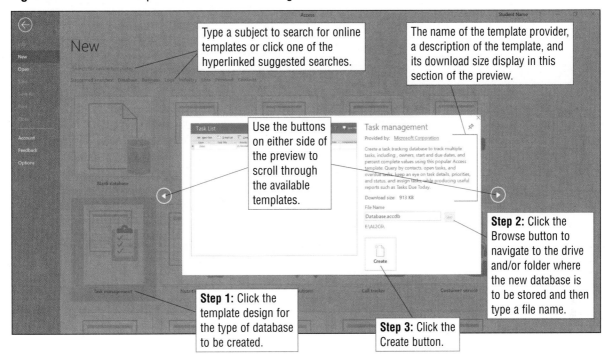

Type a subject to search for online templates or click one of the hyperlinked suggested searches.

The name of the template provider, a description of the template, and its download size display in this section of the preview.

Use the buttons on either side of the preview to scroll through the available templates.

Step 1: Click the template design for the type of database to be created.

Step 2: Click the Browse button to navigate to the drive and/or folder where the new database is to be stored and then type a file name.

Step 3: Click the Create button.

Activity 1a Creating a New Contacts Database Using a Template
Part 1 of 3

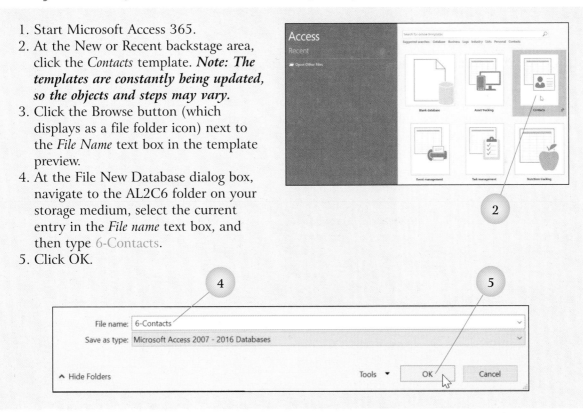

1. Start Microsoft Access 365.
2. At the New or Recent backstage area, click the *Contacts* template. ***Note: The templates are constantly being updated, so the objects and steps may vary.***
3. Click the Browse button (which displays as a file folder icon) next to the *File Name* text box in the template preview.
4. At the File New Database dialog box, navigate to the AL2C6 folder on your storage medium, select the current entry in the *File name* text box, and then type 6-Contacts.
5. Click OK.

6. Click the Create button.

7. The database is created with all the objects from the template loaded into the current window and the Contact List form open.

8. Click the Enable Content button in the message bar.

9. A Welcome to the Contacts Database form appears that provides information on using this database. Click the *Show Welcome when this database is opened* check box to remove the check mark. If the box is left checked, this form will open every time the database is opened.

10. Close the Welcome form. Notice that the Title bar contains the text *Contact Management Database* rather than the name of the database, *6-Contacts*. (How to change the Title bar to display a more descriptive database title will be discussed in Chapter 7.)

11. Review the list of objects that Access has created in the Navigation pane.

12. Double-click to open the Contacts table.

13. Scroll right to view all the fields in the Contacts table and then close the table.

14. Review the Contact List form that is open in the work area and then close it.

Activity 1b Entering and Viewing Data in the Contacts Database

Part 2 of 3

1. With **6-Contacts** open, open the Contact Details form in Form view, and then add the following record to modify the data source file using the form:

First Name	Ariel
Last Name	Grayson
Company	Grayson Accounting Services
Job Title	Accountant
E-mail	ariel@ppi-edu.net **Note: Press the Tab key four times after this field to move to the Street field.**
Street	17399 Windsor Avenue
City	Detroit
State/Province	MI
Zip/Postal Code	48214-3274
Country/Region	USA
Business Phone	800-555-4988
Home Phone	313-555-9684
Mobile Phone	313-555-6811
Fax Number	800-555-3472
Notes	Ariel recommended to RSR by Pat Hynes.

2. Add a picture of the contact using the *Attachments* field by completing the following steps:
   a. Click the Edit Picture button at the top left part of the form.

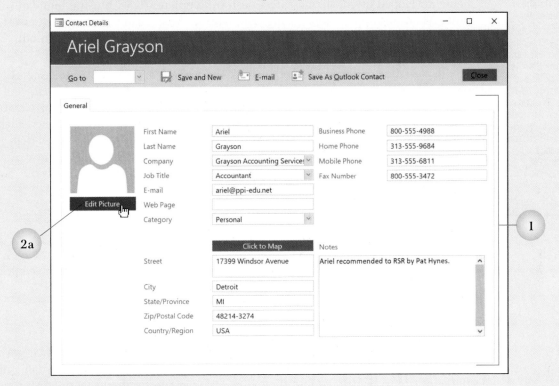

b. At the Attachments dialog box, click the Add button.
c. At the Choose File dialog box, navigate to the AL2C6 folder on your storage medium.
d. Double-click *ArielGrayson.jpg* to add the file to the Attachments dialog box and then click OK.

e. Notice that the picture now displays in the form. If a Word document had been selected in the Attachments dialog box, a Word icon would display.

3. Click the Close button in the upper right corner of the form to close the form. *Note: Do not use the purple Close button right of* Save as Outlook Contact, *as macro errors may occur.*
4. Double-click *Directory* in the Reports group in the Navigation pane. Review the report.
5. Display the report in Print Preview and then print the report.
6. Close Print Preview and then close the Directory report.

Check Your Work

Saving a Database as a Template

Quick Steps

Save Database as Template
1. Open database.
2. Make changes.
3. Click File tab.
4. Click *Save As*.
5. Click Browse button.
6. Change *Save as type* to *Template (*.accdt)*.
7. Type information in Create New Template from This Database dialog box.
8. Click OK.
9. Click OK.

If none of the existing templates is suitable and the same database structure is being developed often, then create a custom template. As with using templates in other Microsoft applications, using templates in Access can save time and effort. To create a database template, either create an entirely new database or modify an existing database based on a template and then save it as a new template.

To save a database as a template, click the File tab and then click the *Save As* option. With *Save Database As* selected in the *File Types* section, click *Template* in the *Database File Types* section and then click the Save As button. Fill in the Create New Template from This Database dialog box, shown in Figure 6.3, and then click OK. Click OK at the Contact Management Database dialog box that states that the template has been saved as [c:]\Users*username*\AppData\Roaming\Microsoft\Templates\Access\TemplateName.accdt. Note the default location in which the templates are stored.

Using a Custom Template

To start a user-created template, click the File tab and then click *New*. At the New backstage area, click Personal. This opens the Personal template area. Double-click the name of a template to open it.

Deleting a Custom Template

To delete a custom template, use the Open dialog box to navigate to [c:]\Users*username*\AppData\Roaming\Microsoft\Templates\Access\. Right-click the name of the template to be deleted and then click *Delete* at the shortcut menu. Click the Cancel button to close the Open dialog box.

Figure 6.3 Create New Template from This Database Dialog Box

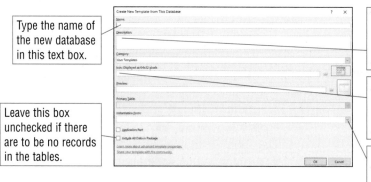

Type the name of the new database in this text box.

Leave this box unchecked if there are to be no records in the tables.

Type a description of the database in this text box, including who created it, when it was created, and other relevant information.

Add an image (png, jpg, jpeg, gif, or bmp) that will represent the new database in the Personal template area of the New backstage area.

Use this option box to specify the form that will display when a database created from this template is opened.

1. With **6-Contacts** open, open the Contact Details form in Design view, click the Close button that contains a close command macro, and then press the Delete key. (Macros will be discussed in Chapter 7.) Save and close the form.
2. Save the revised database as a template by completing the following steps:
   a. Click the File tab.
   b. Click *Save As*.
   c. Click *Template* in the *Database File Types* section.
   d. Click the Save As button.
   e. At the Create New Template from This Database dialog box, enter data as follows:

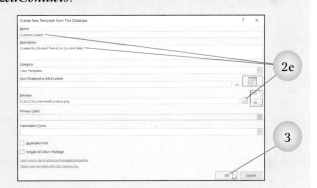

Name	CustomContacts
Description	Created by [Student Name] on [current date]. (Substitute your name for *[Student Name]* and today's date for *[current date]*.)
Preview	Click the Browse button, navigate to the AL2C6 data folder, and then double-click ***CustomizedContacts***.

3. Click OK.
4. Click OK.
5. Close **6-Contacts**.
6. To view the new template, click the File tab and then click Personal. The CustomContacts template created is shown with **CustomizedContacts** as the image. Click the Back button.

Activity 2 Create Objects Using a Template   3 Parts

You will use Application Parts templates to create a series of objects in an existing database. You will also define a form as a template for all the new forms in a database.

Creating Objects Using an Application Parts Template

Application Parts

Access 365 provides templates for prebuilt objects that can be inserted into an existing database using the Application Parts button in the Templates group on the Create tab. The *Quick Start* section of the Application Parts button drop-down list includes the options *Comments*, which creates a table; *Contacts*, which creates a table, a query, forms, and reports; and *Issues*, *Tasks*, and *Users*, each of which creates a table and two forms.

Quick Steps

**Create Objects Using
Application Parts
Template**
1. Open database.
2. Click Create tab.
3. Click Application
   Parts button.
4. Click template.
5. Choose relationship
   options.
6. Add data or modify
   objects as required.
7. Click Create button.

Hint Create your
own Application Parts
template by copying an
object that is reused in
other databases to a
new database and then
saving the database as
a template at the Save
As backstage area.

Creating Objects Using Quick Start

If any of these kinds of tables need to be created in an existing database, consider creating it using the Application Parts template because related objects, such as forms and reports, will be generated automatically. Once the application part is added, any part of an object's design can be modified as needed. To create a group of objects based on a template, click the Create tab and then click the Application Parts button in the Templates group. Click a template in the *Quick Start* section of the drop-down list, as shown in Figure 6.4.

Access opens the Create Relationship Wizard to guide the user through creating the relationship for the new table. Decide in advance of creating the new table what relationship, if any, will exist between the new table and an existing table in the database. At the first Create Relationship dialog box, shown in Figure 6.5, click the first option if the new table will be the "many" table in a one-to-many relationship. Use the drop-down list to choose the "one" table and then click the Next button. If the new table will be the "one" table in a one-to-many relationship, click the second option, use the drop-down list to choose the "many" table, and then click the Next button.

At the second Create Relationship dialog box, enter the settings for the lookup column between the two tables. Choose the field to use to join the tables, choose a sort order if needed, assign the name of the lookup column, and then click the Create button. If the new table will not be related to any of the existing tables in the database, choose the *There is no relationship* option at the first Create Relationship dialog box and then click the Create button.

Figure 6.4 Application Parts Button Drop-Down List

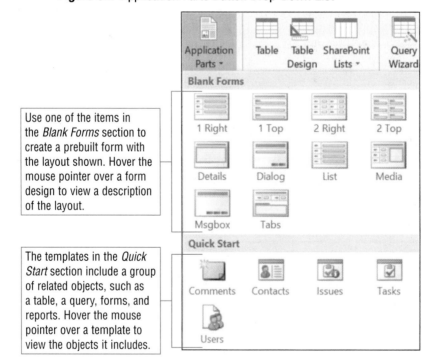

Use one of the items in the *Blank Forms* section to create a prebuilt form with the layout shown. Hover the mouse pointer over a form design to view a description of the layout.

The templates in the *Quick Start* section include a group of related objects, such as a table, a query, forms, and reports. Hover the mouse pointer over a template to view the objects it includes.

Figure 6.5 First Dialog Box in the Create Relationship Wizard

Choose this option if the new table will be the "many" table in a one-to-many relationship.

Choose this option if the new table will be the "one" table in a one-to-many relationship.

Choose this option if the new table will not be related to any existing tables.

Create Relationship ? ✕

Create a simple relationship

You can specify a relationship that would be created when Microsoft Access imports the new template.

One 'Customers' to many 'Contacts'.
Customers ▾

One 'Contacts' to many 'Customers'.
Customers

There is no relationship.

< Back   Next >   Create   Cancel

Activity 2a Creating a Table, a Query, Forms, and Reports Using Application Parts Quick Start Part 1 of 3

1. Open **6-RSRCompServ** and enable the content.
2. Use a template to create a new table, a query, forms, and reports related to contacts by completing the following steps:
   a. Click the Create tab.
   b. Click the Application Parts button in the Templates group and then click *Contacts* in the *Quick Start* section of the drop-down list.

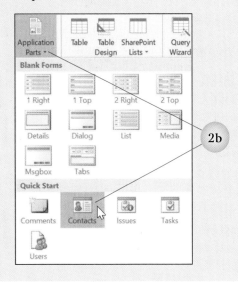

c. The Create Relationship Wizard starts. At the first Create Relationship dialog box, click the *There is no relationship* option and then click the Create button. Access imports a Contacts table; a ContactsExtended query; ContactDetails, ContactDS, and ContactList forms; and ContactAddressBook, ContactList, and ContactPhoneBook reports into the database.

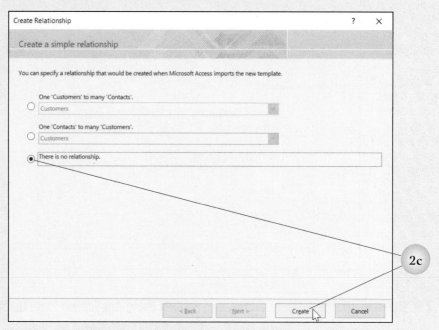

3. Double-click *Contacts* in the Tables group in the Navigation pane. Scroll right to view all the fields in the new table and then close the table.
4. Double-click *ContactDetails* in the Forms group in the Navigation pane.
5. Enter the following record using the ContactDetails form:

First Name	Terry
Last Name	Silver
Job Title	Sales Manager
Company	Cityscape Electronics **Note: Press the Tab key two times after this field to move to the E-mail field.**
E-mail	terry_s@ppi-edu.net
Web Page	(leave blank)
Business Phone	800-555-4968
Fax	800-555-6941
Home Phone	(leave blank)
Mobile Phone	313-555-3442
Address	3700 Woodward Avenue
City	Detroit
State/Province	MI
ZIP/Postal Code	48201-2006
Country/Region	(leave blank)
Notes	(leave blank)

6. Click the Save & Close button at the top right of the form.

7. Open the ContactList report.
8. Display the report in Print Preview and then print it.
9. Close Print Preview and then close the ContactList report.

 Check Your Work

Creating a New Form Using an Application Parts Blank Form

The Application Parts button drop-down list also includes various blank form layouts. Using one of them makes the task of creating a new form easier because the layout has already been defined. The *Blank Forms* section of the Application Parts button drop-down list contains 10 prebuilt blank forms. Most of the forms contain command buttons that perform actions such as saving changes or saving and closing the form. Hover the mouse pointer over a blank form option at the Application Parts button drop-down list to display a description of the form's layout in a ScreenTip. When a blank form option is clicked, Access creates the form object using a predefined form name. For example, if the *1 Right* is clicked, Access creates a form named *SingleOneColumnRightLabels*. Locate the form name in the Navigation pane and then open the form in Layout view or Design view to customize it as needed.

Each Application Parts form has a control layout applied, which means all the form's control objects move and resize together. Remove the control layout to make individual size adjustments. To do this in Design view, select all the control objects and then click the Remove Layout button in the Table group on the Form Design Tools Arrange tab.

Quick Steps

Create Form Using Blank Application Parts Form

1. Click Create tab.
2. Click Application Parts button.
3. Click blank form layout.
4. Add fields to form.
5. Customize form as needed.
6. Save form.

1. With **6-RSRCompServ** open, use an
   Application Parts blank form to create a new
   form for maintaining records in the Parts table by
   completing the following steps:
   a. If necessary, click the Create tab.
   b. Click the Applications Parts button in the
      Templates group and then click *1 Right* in the
      Blank Forms section of the drop-down list. Access
      creates a form named *SingleOneColumnRightLabels*.
   c. If necessary, position the mouse pointer on the
      right border of the Navigation pane until the
      pointer changes to a left-and-right-pointing arrow
      and then drag right to widen the Navigation pane
      until all the object names are visible.
   d. Double-click *SingleOneColumnRightLabels* in the Forms
      group in the Navigation pane.
2. Switch to Layout view.
3. Click to select the *Field1* label control object. Press and
   hold down the Shift key; click to select the *Field2*, *Field3*,
   and *Field4* label control objects, release the Shift key, and
   then press the Delete key.
4. Associate the Parts table with the form and add fields
   from the Parts table by completing the following steps:
   a. If the Field List task pane is not open, click the Add
      Existing Fields button in the Tools group on the Form
      Layout Tools Design tab.
   b. Click the Show all tables hyperlink at the top of the task pane. Skip this step if the
      Field List task pane already displays all the table names in the database.

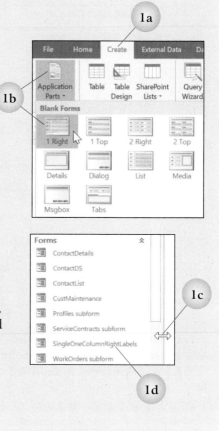

   c. Click the plus symbol
      next to the *Parts* option
      in the Field List task
      pane to expand the list
      and show all the fields in
      the Parts table.
   d. Drag the *PartNo* field to
      the second column in
      the row, as shown at the
      right.
   e. Drag the remaining fields
      (*PartName*, *Supplier*,
      and *Cost*) below
      PartNo, as shown at
      the right.
   f. Close the Field List
      task pane.
5. With the *Cost* field selected, drag up the bottom orange border of the control to decrease
   the height of the control object so the bottom border is directly below the label text.

6. Select the four label control objects and drag the right orange border of the selected control objects right to widen the labels until all the label text is visible.
7. With the four label control objects still selected, click the Form Layout Tools Arrange tab, click the Control Padding button in the Position group, and then click *Wide* at the drop-down list.

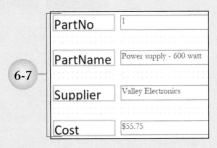

8. Double-click the form title to place the insertion point inside the title text, delete *Form Title*, type Repair Parts, and then press the Enter key.
9. Make formatting changes to the form as follows:
   a. Click the Save button at the top right of the form. Press and hold down the Ctrl key, click the Save & Close button, and then release the Ctrl key. Click the Form Layout Tools Format tab, click the Quick Styles button, and then click *Intense Effect - Blue-Gray, Accent 3* (fourth column, last row).
   b. Select the *Repair Parts* title control object and apply the Blue-Gray, Accent 3 font color (seventh column, first row in the *Theme Colors* section).
   c. Apply the Blue-Gray, Accent 3 font color to the four label control objects.
   d. Apply the Blue-Gray, Accent 3 Lighter 80% background color (seventh column, second row in the *Theme Colors* section) to the four text box control objects adjacent to the labels.
   e. Click the label control object that displays the text *Supplier*. Use the bottom sizing handle to increase the height of the object so all the text displays.

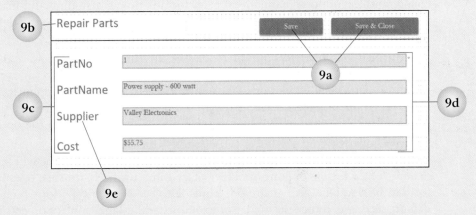

   f. Click in a blank area of the form to deselect the label control object.
10. Click the File tab, click the *Save As* option, click *Save Object As*, and then click the Save As button. Type Parts in the *Save 'SingleOneColumnRightLabels' to* text box at the Save As dialog box and then click OK.
11. Click the Home tab and then switch to Form view.

12. Scroll through a few records in the Parts form and then click the Save & Close button at the top right. Your record may look different than the one shown below.

13. The SingleOneColumnRightLabels form is no longer needed. Click *SingleOneColumnRightLabels* in the Forms group in the Navigation pane and then press the Delete key. Answer Yes to the question *Do you want to permanently delete the form 'SingleOneColumnRightLabels'?*

Check Your Work

Tutorial
Creating a User-Defined Form Template

Setting Form Control Defaults and Creating a User-Defined Form Template

Quick Steps

Create User-Defined Form Template
1. Create new form in Design view.
2. Add control object to form.
3. Format control object.
4. Click More button in Control group.
5. Click *Set Control Defaults.*
6. Repeat Steps 2–5 for each type of control object to be used.
7. Save form, naming it *Normal.*

Hint Delete the Normal form if you want to go back to using the standard default options for control objects in forms.

Using a standard design for all the forms in a database is an effective way of ensuring consistency and portraying a professional image—for example, formatting all the label control objects as 14-point blue text on a gray background. However, changing these options manually in each form is time consuming and may lead to inconsistencies. To save time and ensure consistency, customize the default settings instead.

The *Set Control Defaults* option at the Controls drop-down list allows the user to change the default properties for all the new labels in a form. To do this, open the form in Design view, format one label control object, click the More button in the Controls group, and then click *Set Control Defaults* at the drop-down list. All the new label control objects added to the form will now be formatted with the options that have been defined.

To further customize the database, create a form template that will set the default for each type of control object placed in a form. To do this, open a new form in Design view and create one control object of each type for which a default setting is to be specified. For example, add a label control object, text box control object, command button, and list box, making sure to format each with colors and backgrounds. When each control object is finished, use the *Set Control Defaults* option to change the default settings. Another option is to select all the control objects after the form is finished and then perform one *Set Control Defaults* command. Save the form using the name *Normal.* The Normal form becomes the template for all the new forms in the database. Existing forms retain their initial formatting, unless they are changed manually.

1. With **6-RSRCompServ** open, create a form template for the database by completing the following steps:

   a. Click the Create tab and then click the Form Design button in the Forms group.

   b. Click the Themes button in the Themes group. Notice that the database uses the Organic theme for any new forms. Click in a blank area of the form to close the Themes list.

   c. Click the Label button in the Controls group on the Form Design Tools Design tab, draw a label in the *Detail* section, type Sample Label Text, and then press the Enter key. Do not be concerned with the position and size of the label control object at this time, since this control object will be used only to set new default formatting options.

   d. Click the Form Design Tools Format tab. Apply bold formatting and change the font color to Blue-Gray, Accent 3 (seventh column, first row in the *Theme Colors* section) and the background color to Blue-Gray, Accent 3, Lighter 80% (seventh column, second row in the *Theme Colors* section).

   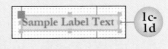

   e. Click the Form Design Tools Design tab, click the Text Box button in the Controls group, and then draw a text box control object in the *Detail* section. Format the text box control object and its associated label control object as follows:

      1) To the label control object, apply bold formatting and the same font and background colors that were applied to the label control object in Step 1d.

      2) Select the text box control object (displays *Unbound*) and apply the same background color applied to the label control object in Step 1d.

   Your text box and combo box numbers may vary.

   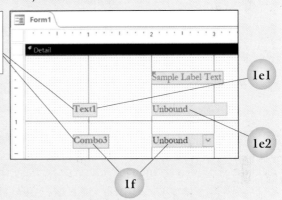

   f. Click the Form Design Tools Design tab, click the Combo Box button in the Controls group, and then draw a combo box in the *Detail* section. If the Combo Box Wizard begins, click Cancel. Format the combo box control objects using the same options applied to the text box control object in Steps 1e1 and 1e2.

   g. Press Ctrl + A to select all the control objects in the form.

   h. Click the Form Design Tools Design tab, click the More button in the Controls group, and then click *Set Control Defaults*.

   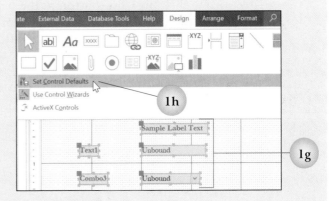

2. Save the form with the name *Normal*.

3. Close the Normal form. Normal becomes the form template for **6-RSRCompServ**. Any new form created will have labels, text boxes, and combo boxes formatted as specified in Step 1.

   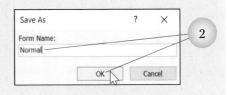

4. Click *WorkOrders* in the Tables group in the Navigation pane, click the Create tab, and then click the Form button in the Forms group. The new WorkOrders form uses the formatting applied to the labels, text boxes, and combo boxes in the Normal form template.

5. With the first record displayed in the WorkOrders form in Form view, click the File tab, click *Print*, and then click *Print Preview*. Click the Columns button in the Page Layout group. Select the current value in the *Width* text box in the *Column Size* section, type 7.5, and then click OK. Close Print Preview.

6. Close the WorkOrders form. Click Yes when prompted to save changes to the design of the form and then click OK to accept the default form name *WorkOrders*.

 Check Your Work

---

Activity 3 Copy Table Structure   1 Part

You will create a new table to store contact information for manufacturer sales representatives by copying an existing table's field names and field properties.

 Tutorial

Copying a Table Structure to a New Table

Quick Steps

Copy Table Structure
1. Select table.
2. Click Copy button.
3. Click Paste button.
4. Type new table name.
5. Click *Structure Only*.
6. Click OK.

💡 **Hint** If you are creating a new table that will be similar to an existing table, save time by copying the existing table's structure and then adding, deleting, and modifying fields in Design view.

Copying a Table Structure to a New Table

Use the copy and paste commands to create a new table that uses the same or similar fields as an existing table. For example, in Activity 3, the structure of the Contacts table will be copied to create a new table for manufacturer contacts that need to be maintained separately from other contact records. Since the fields needed for the manufacturer contact records are the same as those for the existing contact records, base the new table on the existing table.

To copy a table structure, click the existing table name in the Navigation pane, click the Copy button in the Clipboard group on the Home tab, and then click the Paste button in the Clipboard group. When a table has been copied to the Clipboard, clicking the Paste button causes the Paste Table As dialog box to appear, as shown in Figure 6.6. Type the name for the new table in the *Table Name* text box, click *Structure Only* in the *Paste Options* section, and then click OK. Once the table has been created, fields can be added, deleted, and modified as needed.

Figure 6.6 Paste Table As Dialog Box

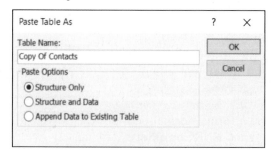

Activity 3 Copying a Table Structure to Create a New Table

1. With **6-RSRCompServ** open, copy the structure of the Contacts table by completing the following steps:
   a. Click *Contacts* in the Tables group in the Navigation pane.
   b. Click the Home tab and then click the Copy button.
   c. Click the Paste button. (Do not click the Paste button arrow.)
   d. At the Paste Table As dialog box, type MfrContacts in the *Table Name* text box and then click *Structure Only* in the *Paste Options* section.
   e. Click OK.

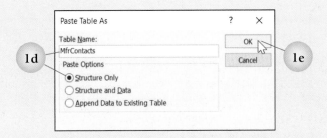

2. Add a description to the table noting that it is based on the Contacts table by completing the following steps:
   a. Right-click MfrContacts in the Navigation Pane.
   b. Click *Table Properties* at the shortcut menu.
   c. Type Based on the Contacts table. in the *Description* text box and then click OK
3. Open the MfrContacts table. Note that the table structure contains the same fields as the Contacts table.
4. Enter the following data in a new record using Datasheet view. Press the Tab key to move past the remaining fields after *ZIP/Postal Code*.

ID	(AutoNumber)
Company	Dell Inc.
Last Name	Haldstadt
First Name	Cari
Email Address	haldstadt@ppi-edu.net
Job Title	Northeast Sales Manager
Business Phone	800-555-9522
Home Phone	(leave blank)
Mobile Phone	800-555-4662
Fax Number	800-555-7781
Address	One Dell Way
City	Round Rock
State/Province	TX
ZIP/Postal Code	78682

5. Close the table.

6. With *MfrContacts* selected in the Tables group in the Navigation pane, click the Create tab and then click the Form button in the Forms group. The form should appear similar to the one shown below.

7. With the form in Layout view, select the label control object and text box control object for the *Attachments* field. Press the Delete key.
8. To move the *Contact Name* field and *File As* field up one spot, select the label and text box control objects for both of these fields and then click the Move Up button in the Move group on the Form Layout Tools Arrange tab.
9. Click the File tab, click the *Print* option, and then click *Print Preview* to display the form in Print Preview. Change to landscape orientation, change the left and right margins to 1 inch, and then close Print Preview.
10. Save the form using the default name *MfrContacts*.
11. Print the selected record and then close the form.

 Check Your Work

Activity 4 **Use Access Tools to Optimize and Document a Database** **5 Parts**

You will use the Table Analyzer Wizard to improve a table's design and the Performance Analyzer to optimize database design and you will split a database by separating tables from queries, forms, and reports. Finally, you will use the Database Documenter to print a report that documents the table structure.

Evaluating a Table Using the Table Analyzer Wizard

Repeating information within a table can result in inconsistencies and wasted storage space. Use the Table Analyzer Wizard to examine a table and determine if duplicate information can be split into smaller related tables to improve the table design. The wizard suggests fields that can be separated into a new table and remain related to the original table with a lookup field. The user can either accept the proposed solutions or modify the suggestions.

Activity 4a uses the Table Analyzer Wizard in a new Parts table. The table was created to store information about the parts commonly used by technicians at RSR Computer Services. Access will examine the table and propose moving the *Supplier* field to a separate table. Making this change will improve the design because several parts records can be associated with the same supplier. In the current table design, the supplier name is typed into a field in each record. Since several parts are associated with each supplier name, the field contains many duplicate entries that take up more storage space than necessary. Furthermore, entering the same data more than once increases the potential for introducing errors in records, which could result in a query that produces incorrect results.

To begin the Table Analyzer Wizard, click the Database Tools tab and then click the Analyze Table button in the Analyze group. This opens the first Table Analyzer Wizard dialog box, shown in Figure 6.7.

Figure 6.7 First Table Analyzer Wizard Dialog Box

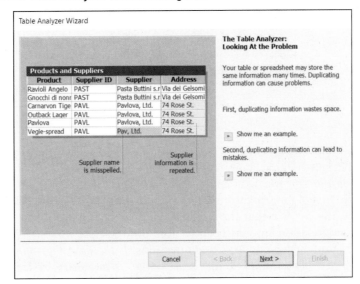

At the first two dialog boxes, review the information about what the Table Analyzer can do to improve the table design. At the third dialog box in the wizard, select the table to be analyzed. At the fourth dialog box, either choose to let the wizard decide which fields to group in the smaller tables or split the tables manually by dragging and dropping fields.

The wizard looks for fields with repetitive data and then suggests a solution. Confirm the grouping of fields and primary key fields in the new tables. At the final step in the wizard, elect to have Access create a query so the fields in the split tables are presented together in a datasheet that resembles the original table.

Activity 4a Splitting a Table Using the Table Analyzer Wizard   Part 1 of 5

1. With **6-RSRCompServ** open, open the Parts table in Datasheet view and review the table structure and data. Notice that the table includes four fields: *PartNo*, *PartName*, *Supplier*, and *Cost*. Also notice that the supplier names are repeated in the *Supplier* field.
2. Close the Parts table.
3. Use the Table Analyzer Wizard to evaluate the Parts table design by completing the following steps:
   a. Click the Database Tools tab.
   b. Click the Analyze Table button in the Analyze group.
   c. Read the information at the first Table Analyzer Wizard dialog box and then click the Next button.
   d. Read the information at the second Table Analyzer Wizard dialog box and then click the Next button.
   e. At the third Table Analyzer Wizard dialog box with *Parts* selected in the *Tables* list box, click the Next button.

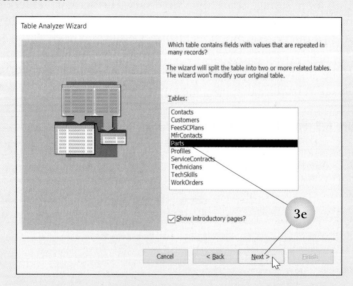

   f. At the fourth Table Analyzer Wizard dialog box with *Yes, let the wizard decide* selected for *Do you want the wizard to decide what fields go in what tables?*, click the Next button.
   g. At the fifth Table Analyzer Wizard dialog box, look at the two tables the wizard is proposing. Notice that the *Supplier* field has been moved to a new table and that a one-to-many relationship has been created between the tables. Access names the new tables *Table1* and *Table2* and asks two questions: *Is the wizard grouping information correctly?* and *What name do you want for each table?* **Note: If necessary, resize the table field list boxes to see all the proposed fields.**

h. The proposed tables have the fields grouped correctly. Double-click the *Table1* Title bar to give the table a new name.

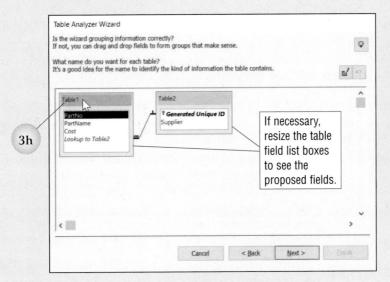

i. Type PartsAndCosts in the *Table Name* text box and then click OK.

j. Click the *Table2* Title bar and then click the Rename Table button near the top right of the dialog box, above the table field list boxes. Type PartsSuppliers in the *Table Name* text box and then press the Enter key.

k. Click the Next button.

l. At the sixth Table Analyzer Wizard dialog box, the unique identifier, or primary key field, for each table is set and/or confirmed. The primary key field is displayed in bold in each table field list box. Notice that the PartsAndCosts table does not have a primary key field defined. Click the *PartNo* field in the PartsAndCosts table field list box and then click the Primary Key button near the top right of the dialog box. Access sets the *PartNo* field as the primary key field, displays a key icon, and applies bold formatting to the field name.

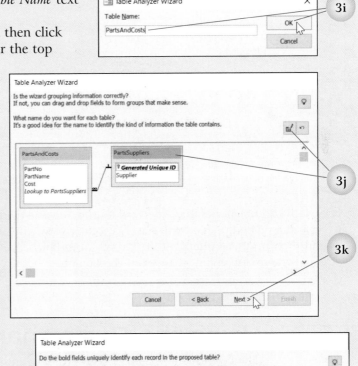

m. Click the Next button.

n. At the last Table Analyzer Wizard dialog box, you can choose to have Access create a query with the original table name that includes the fields from the new tables. Creating the query means that existing forms or reports that were based on the original table will still operate. With *Yes, create the query* selected, click the Finish button. Access renames the original table *Parts_OLD*, creates the query with the name *Parts*, and opens the Parts query results datasheet with the Access Help Task pane.

o. Close the Help task pane.

4. Examine the Parts query datasheet and the object names added in the Navigation pane, including the new tables, PartsAndCosts and PartsSuppliers, along with the original table, Parts_OLD. The Parts query looks just like the original table opened in Step 1, with the exception of the additional field named *Lookup to PartsSuppliers*. The lookup field displays the supplier name, which is also displayed in the original *Supplier* field.

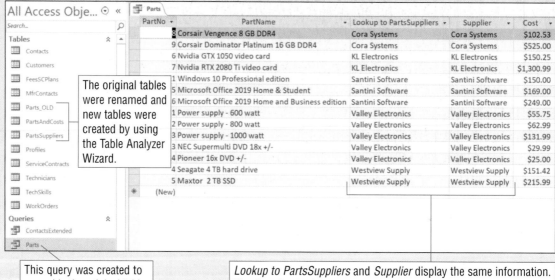

The original tables were renamed and new tables were created by using the Table Analyzer Wizard.

This query was created to resemble the original table.

Lookup to PartsSuppliers and *Supplier* display the same information. The *Supplier* field is deleted from the query in Step 5.

5. Switch to Design view and then delete the *Supplier* field.

6. Save the revised query. Switch to Datasheet view, adjust all the column widths to best fit, and then print the query results datasheet in landscape orientation.

7. Close the query, saving changes to the layout.

Check Your Work

Tutorial

Using the Performance Analyzer

Analyze Performance

Using the Performance Analyzer

The Performance Analyzer can evaluate an individual object, a group of objects, or an entire database for ways that objects can be modified to optimize the use of system resources (such as memory) and improve the speed of data access. If a database seems to run slowly, consider running the tables, queries, forms, reports, or entire database through the Performance Analyzer.

To use the Performance Analyzer, click the Database Tools tab and then click the Analyze Performance button in the Analyze group to open the Performance Analyzer dialog box, shown in Figure 6.8. Select an object type tab, click the check box next to an object to have it analyzed, and then click OK. Select multiple objects or click the Select All button to select all the objects on the

Figure 6.8 Performance Analyzer Dialog Box

Optimize Database Performance
1. Click Database Tools tab.
2. Click Analyze Performance button.
3. Click All Object Types tab.
4. Click Select All button.
5. Click OK.
6. Review *Analysis Results* items.
7. Optimize relevant recommendations or suggestions.
8. Click Close button.

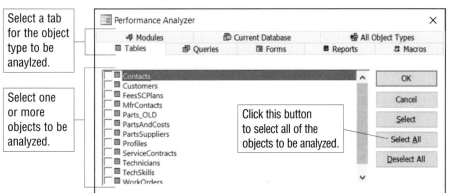

Hint Before running the Performance Analyzer, make sure all the objects to be analyzed are closed. If they are open, the Performance Analyzer will skip them.

current tab for analysis. To evaluate the entire database, click the All Object Types tab and then click the Select All button. Click OK to begin the analysis.

Three types of results are presented to optimize the selected objects: recommendation, suggestion, and idea. Click an item in the *Analysis Results* list box to read a description of the proposed optimization method in the *Analysis Notes* section. Click a recommendation or suggestion in the *Analysis Results* list box and then click the Optimize button to instruct Access to carry out the recommendation or suggestion. Access will modify the object and mark the item as fixed when completed. The Performance Analyzer may provide ideas for improving the design, such as assigning a different data type for a field based on the type of data that has been entered into records or creating relationships between tables that are not already related.

Activity 4b Analyzing a Database to Improve Performance

Part 2 of 5

1. With **6-RSRCompServ** open, use the Performance Analyzer to evaluate the database by completing the following steps:
   a. If necessary, click the Database Tools tab.
   b. Click the Analyze Performance button in the Analyze group.
   c. At the Performance Analyzer dialog box, click the All Object Types tab.
   d. Click the Select All button.
   e. Click OK. The Performance Analyzer displays the name of each object as it is evaluated and then presents a full list of results when the analysis is finished.

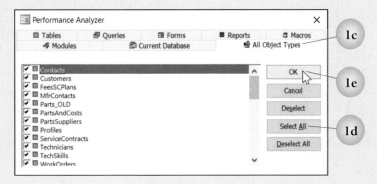

2. Review the items in the *Analysis Results* list box and then optimize a relationship by completing the following steps:

   a. Click the first entry in the *Analysis Results* list box (contains the text *Application: Save your application as an MDE file*) and then read the description of the idea in the *Analysis Notes* section. (Saving an application as an MDE file will be discussed in Chapter 7.)

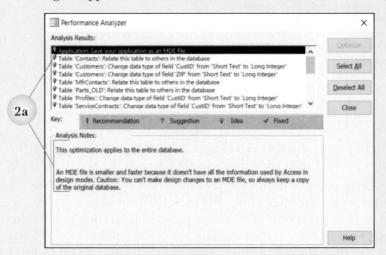

   b. Click the fifth entry in the *Analysis Results* list box (contains the text *Table 'MfrContacts': Relate this table to others in the database*) and then read the description of the idea in the *Analysis Notes* section. The contact information stored in this table is for manufacturer sales representatives and cannot be related to any other tables.

   c. Scroll down the *Analysis Results* list box and then click the item with the green question mark, which identifies a suggestion (contains the text *Table 'WorkOrders': Relate to table 'WorkOrders'*). Read the description of the suggestion in the *Analysis Notes* section. Note that the optimization will benefit the TotalWorkOrders query. This optimization refers to a query that contains a subquery with two levels of calculations. The suggestion is to create a relationship to speed up the query calculations.

   d. Click the Optimize button.

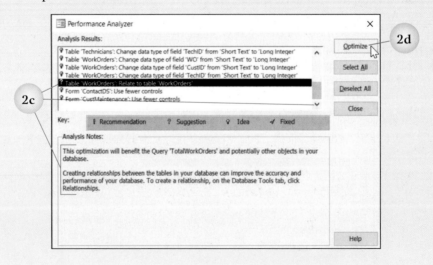

e. Access creates the relationship and changes the question mark next to the item in the *Analysis Results* list box to a check mark (✓). The check mark indicates that the item has been fixed.

f. Click the last item in the *Analysis Results* list box (contains the text *Form 'CustMaintenance': Use fewer controls*) and then read the description of the idea. The idea is to break the form into multiple forms, retaining information that is used often in the existing form. Information viewed less often would be split out into individual forms. To implement this optimization idea would require redesigning the form.

3. Click the Close button to close the Performance Analyzer dialog box.

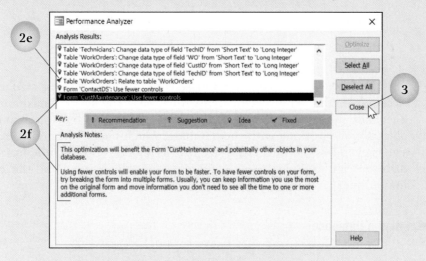

Splitting a Database

Tutorial

Splitting a Database

If a database is placed on a network where multiple users can access it simultaneously, the speed with which the data is accessed may decrease. One way to improve performance is to split the database into two files.

The file containing the tables (called the *back-end database*) is stored in the network share folder, while the file containing the queries, forms, and reports (called the *front-end database*) is stored on individual users' computers. The users can create and/or customize their own queries, forms, and reports to serve their individual purposes. The front-end database contains tables linked to the back-end data, so all the users update a single data source file.

To split an existing database into back-end and front-end databases, Access provides the Database Splitter Wizard. Click the Database Tools tab and then click the Access Database button in the Move Data group to open the first Database Splitter Wizard dialog box, shown in Figure 6.9.

Click the Split Database button to open the Create Back-end Database dialog box and navigate to the drive and/or folder where the database file containing the original tables is to be stored. By default, Access uses the original database file name with *_be* added to the end (before the file extension). Change the file name if required and then click the Split button. Access moves the table objects to the back-end file, creates links to the back-end tables in the front-end file, and then displays a message stating that the database was successfully split.

Quick Steps

Split Database
1. Click Database Tools tab.
2. Click Access Database button.
3. Click Split Database button.
4. If necessary, navigate to drive and/or folder.
5. If necessary, edit file name.
6. Click Split button.
7. Click OK.

Access Database

Hint Consider making a backup copy of the database before splitting the file, in case the database needs to be restored to its original state.

Figure 6.9 First Database Splitter Wizard Dialog Box

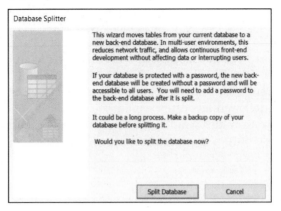

Another reason to split a database is to overcome the file size restriction in Access 365. Database specifications for Access 365 set the maximum file size at 2 gigabytes (GB). This size includes the space needed by Access to open system objects while working with the database, which means the actual maximum file size is even less. However, the size restriction does not include links to external data source files. Splitting a database extends the size beyond the 2-gigabyte limitation.

Activity 4c **Splitting a Database**   Part 3 of 5

1. With **6-RSRCompServ** open, split the database to create back-end and front-end databases by completing the following steps:
   a. If necessary, click the Database Tools tab.
   b. Click the Access Database button in the Move Data group.
   c. At the first Database Splitter Wizard dialog box, click the Split Database button.

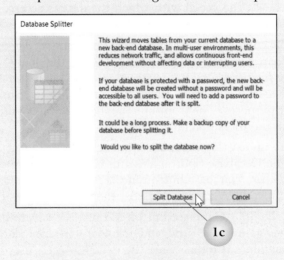

d. At the Create Back-end Database dialog box, navigate to the same folder as the original database (AL2C6). To accept the default file name ***6-RSRCompServ_be*** in the *File name* text box, click the Split button.

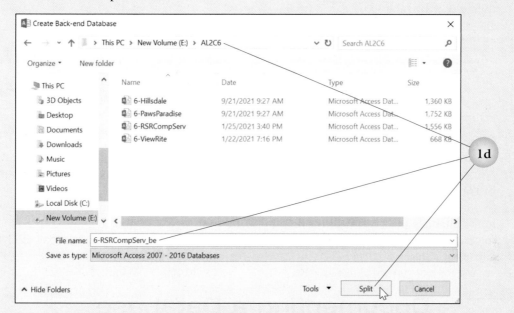

e. Click OK at the Database Splitter message box stating that the database was successfully split.

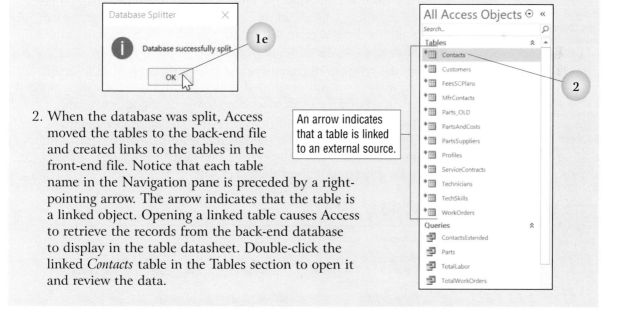

2. When the database was split, Access moved the tables to the back-end file and created links to the tables in the front-end file. Notice that each table name in the Navigation pane is preceded by a right-pointing arrow. The arrow indicates that the table is a linked object. Opening a linked table causes Access to retrieve the records from the back-end database to display in the table datasheet. Double-click the linked *Contacts* table in the Tables section to open it and review the data.

An arrow indicates that a table is linked to an external source.

3. Switch to Design view. Access displays a message stating that the table is a linked table whose design can't be modified and that if you want to add or remove fields or change properties that you will have to do it in the source database the Click the No button and then close the table.

4. Close **6-RSRCompServ**.
5. Open **6-RSRCompServ_be** and enable the content.
6. Notice that the back-end database file contains only the tables. Open the Customers table in Datasheet view and review the data.
7. Switch to Design view. Note that in the back-end database, switching to Design view to make changes does not prompt display of the message box that displayed in Step 3. The message does not display because this database contains the original source table.
8. Close the table.

Tutorial

Documentating a Database

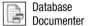

Print Object Documentation
1. Click Database Tools tab.
2. Click Database Documenter button.
3. Click Options button.
4. Choose report options.
5. Click OK.
6. Click object name.
7. Click OK.
8. Print report.
9. Close report.

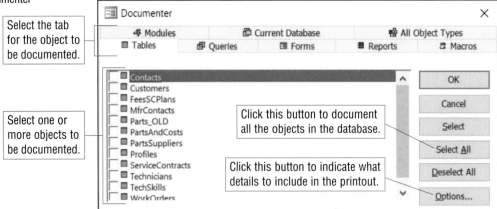

Database Documenter

Documenting a Database

Another Access feature, the Database Documenter feature, prepares and prints a report with details about the definition of a database object. This report can be used as hard-copy documentation of the table structure, including field properties and information regarding a query, form, or report. Relationship diagrams can also be included for all the defined relationships for the table. Relationship options are documented below each relationship diagram.

It is a good idea to store the database documentation report in a secure place. That way, if the data gets corrupted or destroyed, the information needed to manually repair, rebuild, or otherwise recreate the database is available.

Click the Database Tools tab and then click the Database Documenter button in the Analyze group to open the Documenter dialog box, shown in Figure 6.10. Insert a check mark in the check box next to the name of the object the report is to be generated for and then click OK.

Figure 6.10 Documenter Dialog Box

1. With **6-RSRCompServ_be** open, generate a report providing details of the table structure, field properties, and relationships for an individual table by completing the following steps:
   a. Click the Database Tools tab.
   b. In the Analyze group, click the Database Documenter button.
   c. At the Documenter dialog box, click the Options button.
   d. At the Print Table Definition dialog box, click the *Permissions by User and Group* check box in the *Include for Table* section to remove the check mark.
   e. Make sure *Names, Data Types, Sizes, and Properties* is selected in the *Include for Fields* section.
   f. Click *Nothing* in the *Include for Indexes* section.
   g. Click OK.
   h. With Tables the active tab in the Documenter dialog box, click the *PartsSuppliers* check box to insert a check mark.
   i. Click OK.

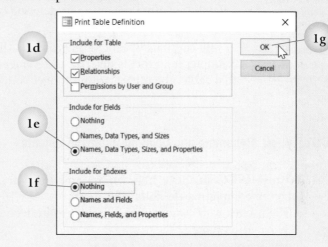

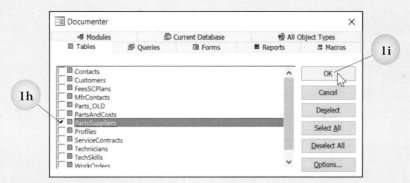

2. Access generates the table definition report and displays it in Print Preview. Print the report.
3. Notice that the Save option is dimmed. A report generated by the Database Documenter cannot be saved, but it can be exported as a PDF, XPS, or one of the other file options found in the Data group on the Print Preview tab. Click the Close Print Preview button in the Close Preview group on the Print Preview tab.

Check Your Work

 Tutorial

Renaming and
Deleting Objects

Renaming and Deleting Objects

As part of managing a database, objects may need to be renamed or deleted within the database file. To do this, right-click the object name in the Navigation pane and then click *Rename* or *Delete* at the shortcut menu. An object may also be deleted from a database by selecting the object in the Navigation pane and then pressing the Delete key. Click Yes at the Microsoft Access message box asking you to confirm deletion of the object. Consider making a backup copy of a database before renaming or deleting objects in case the database needs to be restored to its previous state.

Quick Steps

Rename Object
1. Right-click object in Navigation pane.
2. Click *Rename*.
3. Type new name.
4. Press Enter key.

Delete Object
1. Right-click object in Navigation pane.
2. Click *Delete*.
3. Click Yes.

Be cautious when deleting objects that have dependencies to other objects. For example, if a table is deleted and a query exists that is dependent on fields within that table, the query will no longer run.

Activity 4e **Renaming and Deleting Database Objects**   Part 5 of 5

1. With **6-RSRCompServ_be** open, rename the MfrContacts table by completing the following steps:
   a. Right-click *MfrContacts* in the Tables group in the Navigation pane.
   b. Click *Rename* at the shortcut menu.
   c. Type ManufacturerContacts and then press the Enter key.
2. Delete the original table that was split using the Table Analyzer Wizard in Activity 4a by completing the following steps:
   a. Right-click *Parts_OLD* in the Tables group in the Navigation pane.
   b. Click *Delete* at the shortcut menu.
   c. Click Yes at the Microsoft Access message box asking if you want to delete the table *Parts_OLD*.

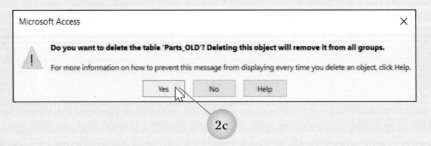

3. Close **6-RSRCompServ_be**.
4. Open **6-RSRCompServ** (the front-end database) and enable the content.

5. Double-click *MfrContacts* in the Tables group in the Navigation pane. Because the table was renamed in Step 1, Access can no longer find the source data. At the Microsoft Access message box stating that the database engine cannot find the input table, click OK. The link will have to be recreated to establish a new connection to the renamed table. (Creating a link to an external table will be discussed in Chapter 8.)

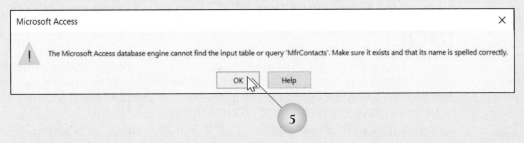

6. Double-click *Parts_OLD* in the Tables group in the Navigation pane. Because this table was deleted in Step 2, the same message appears. Click OK to close the message box.
7. Right-click *Parts_OLD* in the Tables group in the Navigation pane and then click *Delete* at the shortcut menu. Click Yes at the Microsoft Access message box asking if you want to remove the link.

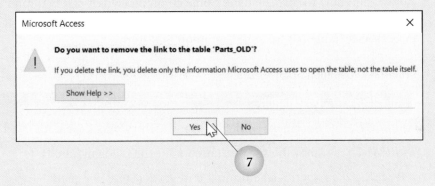

8. Delete the ContactPhoneBook report by completing the following steps:
   a. Click *ContactPhoneBook* in the Reports group in the Navigation pane.
   b. Press the Delete key.
   c. Click Yes at the Microsoft Access message box asking if you want to permanently delete the report.

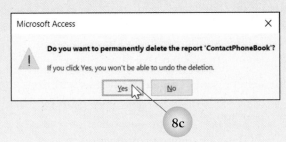

Activity 5 Use Structured Query Language   **1 Part**

You will modify a query using Structured Query Language (SQL).

Using Structured Query Language (SQL)

Hint A typical query may look like this:

SELECT field_1
FROM table_1
WHERE criterion_1;

Note that Access ignores line breaks and uses a semicolon to indicate the end of a statement.

Queries are created to extract data from a database. They are created using the Query Wizard and Query Design buttons in the Queries group on the Create tab. Behind the scenes, Access writes the queries in a computer language called Structured Query Language (SQL). SQL is the computer language used for managing data in relational databases. SQL is fairly easy to learn and understand, but the correct syntax, or order of instructions, is very important.

Each SQL statement begins with a command, such as SELECT, INSERT, and so on. The most common command is SELECT. A SELECT statement is used to retrieve specific data from the database. Each SELECT statement includes several clauses, which give specific instructions about where to find the data, what criteria to use, and how to sort the data. Some common clauses can be found in Table 6.1.

SQL clauses are made up of SQL terms that are comparable to the parts of speech. See Table 6.2 for some common SQL terms.

Table 6.1 Common SQL Clauses

SQL Clause	Description	Required
SELECT	Specifies the fields from which to select the data	Yes
FROM	Lists the tables that contain the fields required	Yes
WHERE	Specifies additional criteria that must be met	No
ORDER BY	Indicates how to sort the data	No

Table 6.2 Common SQL Terms

SQL Term	Comparable Part of Speech	Description	Example
identifier	noun	Used to identify a database object, or content of object such as a field	*Technician.[FName]*
operator	verb or adverb	Represents an action, such as addition (+), or describes how to perform an action (AS)	*SELECT [Technicians] AS [Supervisors]*
constant	noun	A non-changing or null value	*25*
expression	adjective	A combination of identifiers, operators, constants, and functions that produces a result	*>=WorkOrders.[Rate]*

Figures 6.11, 6.12, and 6.13 show a TotalLabor query in SQL, Design, and Datasheet view. Figure 6.11 shows a SELECT statement, beginning with a SELECT clause that selects each of the fields and creates a new field. The FROM clause identifies the table where the fields are located, and the WHERE clause gives additional criteria that must be met in order for the record to be returned.

Figure 6.11 TotalLabor Query SQL View

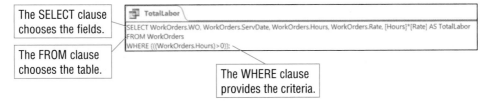

The SELECT clause chooses the fields.

The FROM clause chooses the table.

The WHERE clause provides the criteria.

```
TotalLabor
SELECT WorkOrders.WO, WorkOrders.ServDate, WorkOrders.Hours, WorkOrders.Rate, [Hours]*[Rate] AS TotalLabor
FROM WorkOrders
WHERE (((WorkOrders.Hours)>0));
```

Figure 6.12 TotalLabor Query Design View

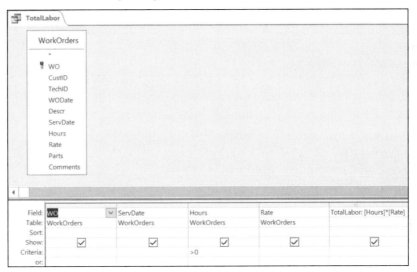

Figure 6.13 TotalLabor Query Datasheet View

Work Order ▾	Service Date ▾	Hours ▾	Rate ▾	Total Labor ▾
65019	Wed, Dec 01 2021	2.50	55.00	$137.50
65020	Thu, Dec 02 2021	1.50	50.00	$75.00
65021	Thu, Dec 02 2021	3.25	55.00	$178.75
65022	Fri, Dec 03 2021	1.50	55.00	$82.50
65023	Sat, Dec 04 2021	1.25	50.00	$62.50
65024	Mon, Dec 06 2021	1.00	45.00	$45.00
65025	Mon, Dec 06 2021	1.50	50.00	$75.00
65026	Tue, Dec 07 2021	0.75	50.00	$37.50

1. With **6-RSRCompServ** open, modify the TotalLabor query to add the customer's ID using SQL view by completing the following steps:

   a. Double-click the *TotalLabor* query in the Navigation pane.

   b. Click the View option arrow in the Views group on the Home tab and then click *SQL View*.

   c. Click directly after the first comma in the SELECT statement.

   d. Press the space bar and then type WorkOrders. CustID,. The first line of code should now read *SELECT WorkOrders.WO, WorkOrders.CustID, WorkOrders.ServDate, WorkOrders.Hours, WorkOrders.Rate, [Hours]*[Rate] AS TotalLabor*

2. Click the Run button. The customer's ID appears between the work order number and the service date.

3. Save the query and then close it.

4. Close **6-RSRCompServ**.

Chapter Summary

- Access provides professionally designed database templates that include predefined tables, queries, forms, and reports for use in creating a new database.

- Choose a database template from the sample templates stored on your computer or download a database template from Office.com.

- Predefined table and related object templates for Comments; Contacts; and Issues, Tasks, and Users are available in the *Quick Start* section of the Application Parts button drop-down list in the Templates group on the Create tab.

- The *Blank Forms* section of the Application Parts button drop-down list contains 10 prebuilt blank forms. Most of these forms include command buttons that perform actions such as saving changes and saving and closing the form.

- Define a custom form template by creating a form named *Normal* that includes a sample of each control object with the formatting options required for future forms. Select all the controls and use the *Set Control Defaults* option to save the new settings.

- When a table has been copied to the Clipboard from the Navigation pane, clicking the Paste button causes the Paste Table As dialog box to open. At this dialog box, choose one of the three options: *Structure Only*, *Structure and Data*, or *Append Data to Existing Table*.

- Use the Table Analyzer Wizard to evaluate a table for repeated data and determine if the table can be split into smaller related tables.
- Use the Performance Analyzer to evaluate an individual object, a group of objects, or an entire database for ways to optimize the use of system resources and improve the speed of data access.
- The Performance Analyzer provides three types of results in the *Analysis Results* list box: recommendation, suggestion, and idea.
- Click a recommendation or suggestion in the *Analysis Results* list box and then click the Optimize button to instruct Access to carry out the recommendation or suggestion.
- A database can be split into two individual files—a back-end database and a front-end database—to improve performance for a multiuser database or overcome the maximum database file size restriction.
- To split a database using the Database Splitter Wizard, begin by clicking the Access Database button in the Move Data group on the Database Tools tab.
- Use the Database Documenter feature to print a hard-copy report with details about the definition of a database object, query, form, or report or the structure of a table.
- Rename an object by right-clicking the object name in the Navigation pane, clicking *Rename* at the shortcut menu, typing the new name, and then pressing the Enter key.
- Delete an object by right-clicking the object name in the Navigation pane, clicking *Delete* at the shortcut menu, and then clicking Yes at the message box asking you to confirm deletion of the object.
- Use SQL view to modify queries using common SQL clauses such as SELECT, FROM, and WHERE.

Commands Review

FEATURE	RIBBON TAB, GROUP	BUTTON	KEYBOARD SHORTCUT
Application Parts template	Create, Templates		
Database Documenter	Database Tools, Analyze		
Paste Table As	Home, Clipboard		Ctrl + V
Performance Analyzer	Database Tools, Analyze		
split database	Database Tools, Move Data		
Table Analyzer Wizard	Database Tools, Analyze		

Microsoft®
Access®

Automating, Customizing, and Securing Access

Performance Objectives

Upon successful completion of Chapter 7, you will be able to:

1 Create, run, edit, and delete a macro

2 Create a command button to run a macro

3 View an embedded macro in the Property Sheet task pane

4 Convert a macro to Visual Basic for Applications

5 Create and edit a Navigation form

6 Customize database startup options

7 Limit access to options in ribbons and menus

8 Customize the Navigation pane

9 Define error-checking options

10 Import and export customized settings

11 Customize the ribbon and the Quick Access Toolbar

Macros are used to automate repetitive tasks and to store these actions so that they can be executed by clicking a button in a form. A Navigation form is used as a menu to provide an interface between the user and objects within the database file. In this chapter, you will learn how to automate a database using macros and a Navigation form. You will also learn how to customize the Access environment.

Data Files

Before beginning chapter work, copy the AL2C7 folder to your storage medium and then make AL2C7 the active folder.

The online course includes additional training and assessment resources.

Tutorial

Creating and Running a Macro

Hint To create a complex macro, begin by working through the steps to be saved. Write down all the parameters before attempting to record the macro.

Creating and Running a Macro

A macro is a series of instructions stored in sequence that can be recalled and carried out as needed. A macro is generally created for a task that never varies and that is frequently repeated. For example, a macro could be created to open a query, form, or report. The macro object stores a series of instructions (called *actions*) in the sequence in which they are to be performed. Macros appear as objects within the Navigation pane. A macro that opens a query, form, or report can be run by double-clicking the macro name in the Navigation pane.

Not all macros can be run by this method, however. Some need to be assigned to buttons that are clicked in order to run them. For example, a user can create a macro in a form that automates the process of searching through the records for a specific client's last name. The macro will contain two instructions: to move to the field in which the last name is stored and to open the Find and Replace dialog box. In general, a macro that does not include an instruction containing the query, form, or report name must be assigned to a button. Otherwise, an error message will appear stating that the object is not currently selected or is not in the active view.

To create a macro, click the Create tab and then click the Macro button in the Macros & Code group. This opens the Macro Builder window, shown in Figure 7.1. Click the arrow at the right side of the *Add New Action* option box and then click an instruction at the drop-down list. As an alternative, add an action using the Action Catalog task pane, as described in Figure 7.1.

Figure 7.1 Macro Builder Window

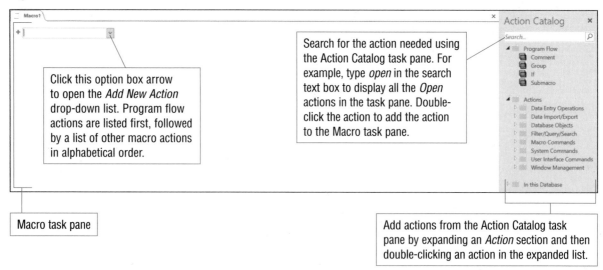

Quick Steps

Create Macro
1. Click Create tab.
2. Click Macro button.
3. Click *Add New Action* option box arrow.
4. Click action.
5. Enter arguments as required in Macro task pane.
6. Click Save button.
7. Type name for macro.
8. Press Enter key.
9. Repeat Steps 3–6 as needed.

Run Macro
Double-click macro name in Navigation pane.
OR
1. Right-click macro name.
2. Click *Run.*
OR
With macro open in Design view, click Run button.

Hint Hover the mouse pointer over the entry box for an argument to reveal a description of the argument and the available parameters in a ScreenTip.

 Run

 Macro

Each new action entered into the Macro task pane is associated with a set of arguments that displays once the action has been added. Similar to the field properties displayed in Table Design view, the arguments displayed in the Macro task pane vary depending on the active action that has been expanded. Figure 7.2 displays the action arguments for the OpenForm action.

The OpenForm action is used to open a form. Within the Macro task pane, specify the name of the form to open and the view in which the form is to be presented. Choose to open the form in Form view, Datasheet view, Layout view, Design view, or Print Preview. Use the Filter Name or Where Condition argument to restrict the records displayed in the report.

The Data Mode argument is used to place editing restrictions on records while the form is open. Choose to open the form in Add mode to allow only adding new records (users cannot view existing records); Edit mode to allow adding, editing, and deleting records; or Read Only mode to allow viewing records. The Window Mode argument is used to instruct Access to open the form in Normal mode (the way forms are normally viewed in the work area), Hidden mode (form is hidden), Icon mode (the form opens minimized), or Dialog mode (the form opens in a separate window similar to a dialog box).

To create a macro that performs multiple actions, use the *Add New Action* option box to add the second action below the first action. Access executes the actions in the order they appear in the Macro task pane. Activity 1a demonstrates how to create a macro with multiple actions that will instruct Access to open a form, make active a control within the form, and then open the Find and Replace dialog box to search for a record.

The GoToControl action is used to make active a control within a form or report, and the RunMenuCommand action is used to execute an Access command. For each action, a single argument specifies the name of the control to move to and the name of the command to run.

As actions are added to the Macro task pane, expand and collapse them as needed. When several actions have been added to the Macro task pane, collapsing them allows the user to focus on the current action being edited.

Figure 7.2 Macro Task Pane with Action Arguments for the OpenForm Action

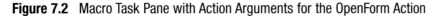

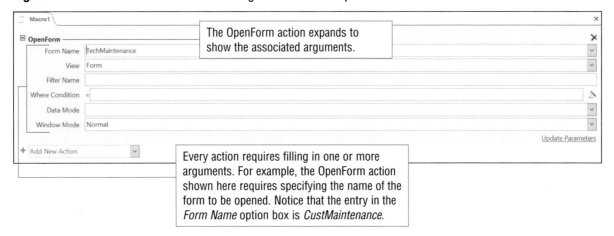

The OpenForm action expands to show the associated arguments.

Every action requires filling in one or more arguments. For example, the OpenForm action shown here requires specifying the name of the form to be opened. Notice that the entry in the *Form Name* option box is *CustMaintenance.*

1. Open **7-RSRCompServ** and enable the content. Create a macro to open the
   TechMaintenance form by completing the following steps:
   a. Click the Create tab.
   b. Click the Macro button in the Macros & Code group.
   c. At the Macro task pane, click the *Add New Action*
      option box arrow, scroll down the list, and then click
      OpenForm. Access adds the action and the arguments
      associated with the action. Most actions require at
      least one argument.
   d. Click the *Form Name* option box arrow in
      the *OpenForm* action section and then click
      TechMaintenance at the drop-down list.

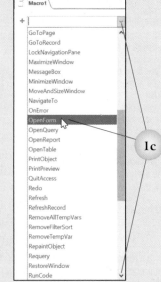

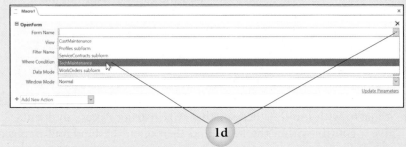

2. Add additional instructions to move to the field for the technician's last name and then
   open the Find and Replace dialog box by completing the following steps:
   a. Click the *Add New Action* option box arrow, scroll down the list, and then click
      GoToControl. Notice that only one action argument is required for the GoToControl
      action.
   b. With the insertion point in the *Control Name* text box in the GoToControl action,
      type LName. (Entering the name of the field that is to be made active in the form is
      required.)

   c. Click the *Add New Action* option box arrow, scroll down the list, and then click
      RunMenuCommand.

d. Click the *Command* option box arrow, scroll down the list, and then click *Find*.

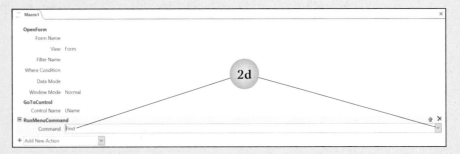

3. Click the Save button on the Quick Access Toolbar, type FormFindTech in the *Macro Name* text box at the Save As dialog box, and then click OK.

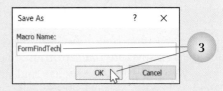

4. Click the Run button (which displays as a red exclamation point) in the Tools group on the Macro Tools Design tab. The Run button instructs Access to carry out the instructions in the macro. The TechMaintenance form opens, the *Last Name* field is active, and the Find and Replace dialog box appears with the last name of the first technician entered in the *Find What* text box. Type Eastman and then click the Find Next button. ***Note: If the Find and Replace dialog box overlaps the* Last Name *field in the form, drag the dialog box to the bottom or right edge of the work area.***

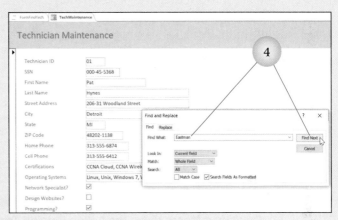

5. Access moves to record 10. Close the Find and Replace dialog box. Notice that the last name text is selected in the form. Read the data displayed in the form for the technician named Kelsey Eastman.

6. Close the form.

7. At the Macro Builder window, click the Close button. This closes the FormFindTech macro. Notice that a Macros group has been added to the Navigation pane and that FormFindTech appears as an object below this new group name. You may need to scroll down the Navigation pane to view the Macros group.

A macro can also be created by dragging and dropping an object name from the Navigation pane into the *Add New Action* option box in a Macro task pane. By default, Access creates an OpenTable, OpenQuery, OpenForm, or OpenReport action depending on the object that was dragged to the window. The object name is also automatically entered in the Object Name argument list box.

Activity 1b Creating a Macro by Dragging and Dropping an Object

1. With **7-RSRCompServ** open, use the drag and drop method to create a macro to open the CustMaintenance form by completing the following steps:
   a. Click the Create tab.
   b. Click the Macro button in the Macros & Code group.
   c. Position the mouse pointer on the CustMaintenance form name in the Navigation pane, click and hold down the left mouse button, drag the object name into the *Add New Action* option box in the Macro task pane, and then release the mouse button. Access inserts an OpenForm action with *CustMaintenance* entered in the *Form Name* option box.

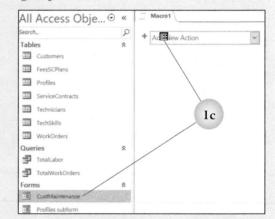

2. Click the Save button, type FormCustMaint, and then click OK.
3. Click the Run button in the Tools group on the Macro Tools Design tab.
4. Close the form.
5. Close the macro by clicking the Close button.

Activity 1c Creating a Macro Using the Action Catalog Task Pane

1. With **7-RSRCompServ** open, use the Action Catalog task pane to create a macro to find a record in a form by making active the home telephone field and then opening the Find and Replace dialog box by completing the following steps. *Note: This macro will be assigned to a button in Activities 1e and 1f. An error will occur if you try to run the macro by double-clicking it in the Navigation pane.*
   a. Click the Create tab.
   b. Click the Macro button.
   c. Click the expand button (which displays as a white triangle) next to *Database Objects* in the *Actions* list in the Action Catalog task pane to expand the category and display the actions available for changing controls or objects in the database. *Note: If the Action Catalog task pane was accidentally closed and does not redisplay in the new Macro Builder window, restore the task pane by clicking the Action Catalog button in the Show/Hide group on the Macro Tools Design tab.*

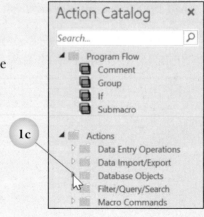

d. Double-click *GoToControl* at the expanded *Database Objects* actions list in the Action Catalog task pane to add the action to the Macro task pane.

e. With the insertion point positioned in the *Control Name* text box in the Macro task pane, type HPhone.

f. Click in the search text box at the top of the Action Catalog task pane and then type Run. As this text is typed, Access displays available actions in the Action Catalog task pane that begin with the same text.

g. Double-click *RunMenuCommand* in the *Macro Commands* list in the Action Catalog task pane.

h. With the insertion point positioned in the *Command* text box in the Macro task pane, type Find.

i. Click the Save button on the Quick Access Toolbar.

j. At the Save As dialog box, type HPhoneFind and then click OK.

k. Close the HPhoneFind macro.

2. Use the Action Catalog to create a macro to close the current database and exit Access by completing the following steps:

a. Click the Create tab and then click the Macro button.

b. Click the box right of the search text box in the Action Catalog task pane to clear *Run* from the search text box and redisplay all the Action Catalog categories.

c. Click the expand button (which displays as a white triangle) next to *System Commands* in the *Actions* list in the Action Catalog task pane.

d. Double-click *QuitAccess* in the expanded *System Commands* actions list.

e. With *Save All* the default argument in the *Options* option box, click the Save button on the Quick Access Toolbar.

f. Type ExitRSRdb at the Save As dialog box and then click OK.

3. Close the ExitRSRdb macro.

4. Double-click *ExitRSRdb* in the Navigation pane.

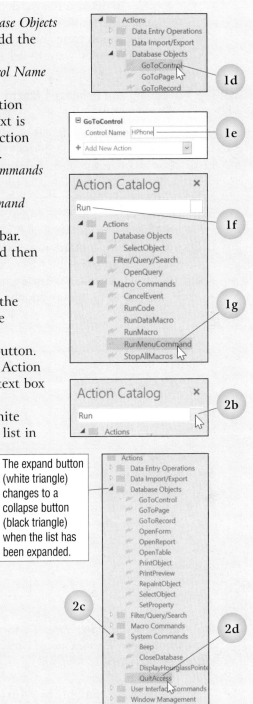

The expand button (white triangle) changes to a collapse button (black triangle) when the list has been expanded.

 Tutorial

Editing a Macro

Editing and Deleting a Macro

To edit a macro, right-click the macro name in the Navigation pane and then click *Design View* at the shortcut menu or right-click the macro name and press Ctrl + Enter. The macro opens in the Macro Builder window. Edit an action and/ or its arguments, insert new actions, and/or delete actions as required. Save the revised macro and close the Macro Builder window when finished.

To delete a macro, right-click the macro name in the Navigation pane and then click *Delete* at the shortcut menu. A macro can also be deleted by selecting it in the Navigation pane and clicking the Delete key. At the Microsoft Access dialog box asking if you want to delete the macro, click Yes.

Activity 1d Editing and Deleting a Macro

1. Open **7-RSRCompServ**. It has been decided that the macro to find a technician record will begin with the TechMaintenance form already opened. This means that the first macro instruction to open the TechMaintenance form in the FormFindTech macro needs to be deleted. To do this, complete the following steps:
   a. If necessary, scroll down the Navigation pane to view the macro object names.
   b. Right-click *FormFindTech* in the Macros group in the Navigation pane and then click *Design View* at the shortcut menu. The macro opens in the Macro Builder window.
   c. Position the mouse pointer over the *OpenForm* action in the Macro task pane. As an action is pointed to in the Macro task pane, Access displays a collapse indicator at the left of the action to allow the arguments to be collapsed; a green arrow at the right to allow the action to be moved; and a black *X* to delete the action.

   d. Click the black *X* at the right side of the Macro task pane across from *OpenForm*. The action is removed from the Macro task pane. ***Note: If the buttons at the right side of the Macro task pane disappear as you move the mouse to the right, move the pointer up so it is on the same line as* OpenForm *to redisplay the buttons.***

2. Save the revised macro. ***Note: This macro can no longer be run by double-clicking it in the Navigation pane. As mentioned earlier, an error will occur. The macro will be assigned to a button in Activities 1e and 1f.***
3. Close the macro.
4. The revised macro contains two instructions that activate the *LName* control and then open the Find and Replace dialog box. This macro can be used in any form that contains a field named *LName*; therefore, the macro should be renamed. To do this, right-click *FormFindTech* in the Navigation pane, click *Rename* at the shortcut menu, type LNameFind, and then press the Enter key.

5. Delete the FormCustMaint macro by completing the following steps:
   a. Right-click *FormCustMaint* in the Navigation pane and then click *Delete* at the shortcut menu.
   b. At the Microsoft Access dialog box asking if you want to delete the macro, click Yes.

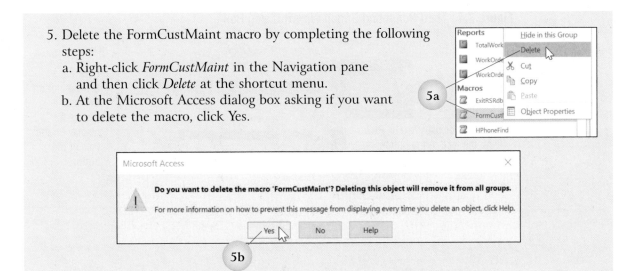

Creating a Command Button to Run a Macro

A macro can be assigned to a button in a form so it can be executed with a single mouse click. This method of running macros makes them more accessible and efficient.

To add a button to a form for running a macro, open the form in Design view. Click the Button button in the Controls group on the Form Design Tools Design tab and then drag to create a button in any form section. When the mouse is released, the Command Button Wizard launches (if the Use Control Wizards feature is active). Recall from Chapter 4 that the Use Control Wizards button displays with a gray background when the feature is activen.

At the first Command Button Wizard dialog box, shown in Figure 7.3, begin by choosing the type of command to assign to the button. Click *Miscellaneous* in the *Categories* list box, click *Run Macro* in the *Actions* list box, and then click the Next button. At the second Command Button Wizard dialog box, choose the name of the macro to assign to the button.

Figure 7.3 First Command Button Wizard Dialog Box

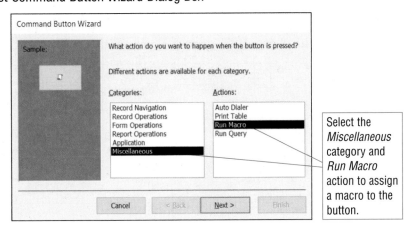

Figure 7.4 Third Command Button Wizard Dialog Box

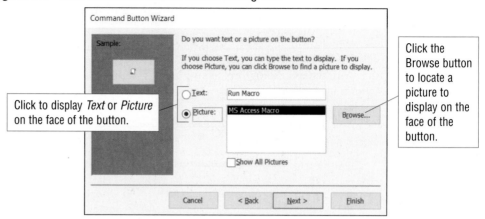

Click to display *Text* or *Picture* on the face of the button.

Click the Browse button to locate a picture to display on the face of the button.

At the third dialog box, shown in Figure 7.4, specify text or a picture to display on the face of the button. When text is entered or a picture file is selected, the button in the *Sample* section of the dialog box updates to show how the button will appear. At the last Command Button Wizard dialog box, assign a name to associate with the command button and then click the Finish button.

Activity 1e Creating a Button and Assigning a Macro to a Button in a Form   Part 5 of 8

1. With **7-RSRCompServ** open, create a command button to run the macro to locate a technician record by last name in the TechMaintenance form by completing the following steps:
   a. Open the TechMaintenance form in Design view. To make room for the new button in the *Form Header* section, click to select the control object with the title text and then drag the right middle sizing handle to the left until the right edge is at approximately the 3.5-inch mark on the horizontal ruler.
   b. By default, the Use Control Wizards button is toggled on in the Controls group. Click the More button in the Controls group on the Form Design Tools Design tab. If the Use Control Wizards button displays with a gray background, the feature is active. If the button is gray, click in a blank area of the form to close the expanded Controls group. If the feature is not active (displays with a white background), click the Use Control Wizards button to turn on the feature.
   c. Click the Button button in the Controls group.
   d. Position the crosshairs with the button icon attached in the *Form Header* section, click and hold down the left mouse button, drag to create a button of the approximate height and width shown below, and then release the mouse button.

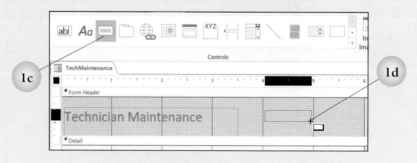

e. At the first Command Button Wizard dialog box, click *Miscellaneous* in the *Categories* list box.
f. Click *Run Macro* in the *Actions* list box and then click the Next button.

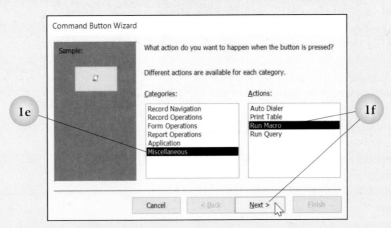

g. At the second Command Button Wizard dialog box, click *LNameFind* in the *What macro would you like the command button to run?* list box and then click the Next button.

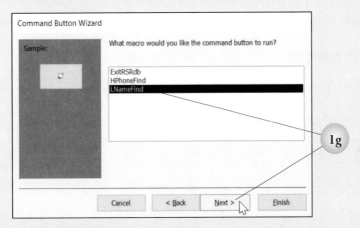

h. At the third Command Button Wizard dialog box, click the *Text* option.
 i. Select the current text in the *Text* text box, type Find Technician by Last Name, and then click the Next button.

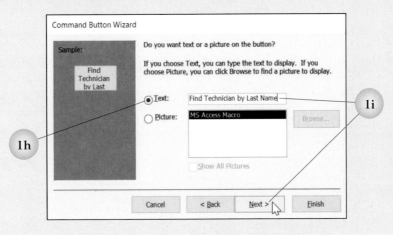

j. With *Command##* (where ## is the number of the command button) already selected in the *What do you want to name the button?* text box, type FindTechRec and then click the Finish button. Access automatically resizes the width of the button to accommodate the text to be displayed on its face.

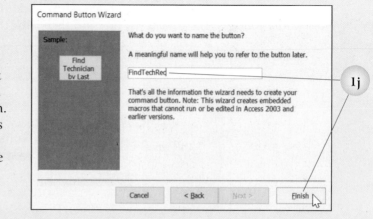

2. Save the revised form.
3. Switch to Form view.
4. Click the Find Technician by Last Name button to run the macro.
5. Type Colacci in the *Find What* text box at the Find and Replace dialog box and then press the Enter key or click the Find Next button. Access makes record 9 active.

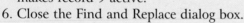

6. Close the Find and Replace dialog box.
7. Close the form.

Activity 1f Creating Two Command Buttons and Assigning Macros to the Buttons

Part 6 of 8

1. With **7-RSRCompServ** open, create a command button to run the macro to find a record by the home telephone number field in the CustMaintenance form by completing the following steps:
   a. Open the CustMaintenance form in Design view.
   b. Click the Button button in the Controls group on the Form Design Tools Design tab.
   c. Position the crosshairs with the button icon attached in the *Detail* section below the *Service Contract?* label control object, click and hold down the left mouse button, drag to create a button of the approximate height and width shown at the right, and then release the mouse button.

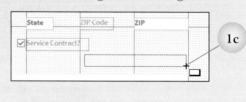

   d. Click *Miscellaneous* in the *Categories* list box, click *Run Macro* in the *Actions* list box, and then click the Next button.
   e. Click *HPhoneFind* and then click the Next button.
   f. Click *Text*, select the current text in the *Text* text box, type Find Customer by Home Phone, and then click the Next button.

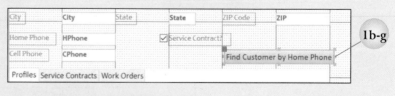

   g. Type FindByPhone and then click the Finish button.

2. Save the revised form.
3. If necessary, move the tab control group down so the top of the object is at the 2.25-inch mark on the vertical ruler. Create a second button below the button created in Step 1 to run the macro to find a record by the last name by completing steps similar to those in Steps 1b through 1g and using the following additional information:
    - Select *LNameFind* as the macro to assign to the button.
    - Display the text *Find Customer by Last Name* on the button.
    - Name the button *FindByLName*.
4. Select both buttons and apply the Intense Effect - Blue, Accent 1 quick style (second column, last row in the *Theme Styles* gallery of the Quick Styles button). Resize, align, and then position the two buttons as shown below.

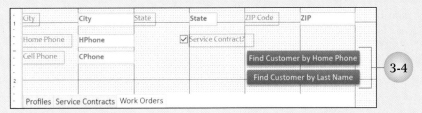

5. Save the revised form.
6. Switch to Form view.
7. Click the Find Customer by Home Phone button, type 313-555-7486 at the Find and Replace dialog box, and then press the Enter key. Close the dialog box and review the record for Customer ID 1025.
8. Click the Find Customer by Last Name button, type Antone at the Find and Replace dialog box, and then press the Enter key. Close the dialog box and review the record for Customer ID 1075.
9. Close the CustMaintenance form.

Viewing an Embedded Macro

Quick Steps
View Macro Code for Command Button
1. Open form in Design view.
2. Click to select command button.
3. Display Property Sheet task pane.
4. Click Event tab.
5. Click Build button in *On Click* property box.

The Command Button Wizard used in Activities 1e and 1f created an embedded macro in the Property Sheet task pane for the button. An embedded macro is a macro stored within a form, report, or control that is run when a specific event occurs, such as clicking a button.

Embedded macros are objects that are not seen in the Navigation pane. To view an embedded macro, open the command button's Property Sheet task pane and then click the Event tab. *[Embedded Macro]* will be visible in the *On Click* property box. Click the Build button (which displays three dots) at the right side of the *On Click* property box to open the Macro Builder window with the macro actions displayed. The macro and macro actions were created by the Command Button Wizard.

To delete an embedded macro, open the Property Sheet task pane, select *[Embedded Macro]* in the *On Click* property box, and then press the Delete key. Close the Property Sheet task pane and then save the form.

1. With **7-RSRCompServ** open, view the macro actions embedded in a command button when the Command Button Wizard is used by completing the following steps:
   a. Open the CustMaintenance form in Design view.
   b. Click to select the Find Customer by Home Phone command button.
   c. Click the Property Sheet button in the Tools group on the Form Design Tools Design tab or press the F4 function key.
   d. Click the Event tab in the Command Button Property Sheet task pane.
   e. Notice that the text in the *On Click* property box reads *[Embedded Macro]*.
   f. Click the Build button (which displays three dots) in the *On Click* property box. When the Build button is clicked, Access opens the Macro Builder window for the macro embedded in the command button.
   g. Notice that the name of the macro created by the Command Button Wizard is *CustMaintenance: FindByPhone: On Click*. The macro's name is made up of the form name, followed by the button name with which the macro is associated, and then the event that causes the macro to run (*On Click*).
   h. Review the macro actions in the Macro task pane. Notice that the macro action is *RunMacro* and that the name of the macro selected at the second Command Button Wizard dialog box, *HPhoneFind*, appears in the *Macro Name* option box.
   i. Click the Close button in the Close group on the Macro Tools Design tab.

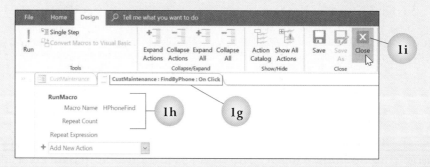

2. With the Property Sheet task pane still open, click to select the Find Customer by Last Name command button and then click the Build button in the *On Click* property box on the Event tab in the Property Sheet task pane.

3. Review the embedded macro name and macro actions in the Macro task pane and then click the Close button in the Close group on the Macro Tools Design tab.

4. Close the form.

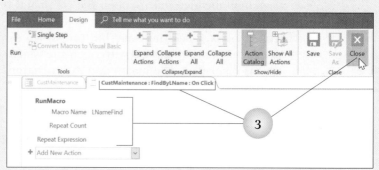

Converting Macros to Visual Basic for Applications

Creating and using macros enables the user to add automation and functionality within Access without having to learn how to write programming code. In the Microsoft Office suite, Visual Basic for Applications (VBA) is the programming language used to build custom applications that operate within Word, Excel, PowerPoint, and Access. The macros used so far in this chapter have been simple enough that they do not need VBA programming. However, when automating more complex tasks, a developer may prefer to write a program using VBA.

A quick method for starting a VBA program is to create a macro and then convert it to VBA code. To do this, open the macro in the Macro Builder window and then click the Convert Macros to Visual Basic button in the Tools group on the Macro Tools Design tab. Access opens a Microsoft Visual Basic window with the VBA code for the macro. When converting an embedded macro, Access also changes the property box in the Property Sheet task pane for the form, report, or control to run the VBA procedure instead of the macro.

Activity 1h Converting a Macro to Visual Basic for Applications   Part 8 of 8

1. With **7-RSRCompServ** open, convert a macro to Visual Basic for Applications (VBA) by completing the following steps:
   a. Right-click *HPhoneFind* in the Macros group in the Navigation pane and then click *Design View* at the shortcut menu. The Macro Builder window opens with the macro actions and arguments for the HPhoneFind macro.
   b. Click the Convert Macros to Visual Basic button in the Tools group on the Macro Tools Design tab.

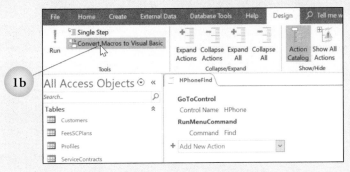

   c. At the Convert macro: HPhoneFind dialog box with the *Add error handling to generated functions* and *Include macro comments* check boxes selected, click the Convert button.
   d. At the *Convert macros to Visual Basic* message box alerting that the conversion has finished, click OK.

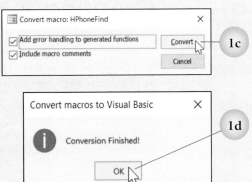

2. Access opens a Microsoft Visual Basic for Applications window when the macro is converted and displays the converted event procedure below the expanded *Modules* folder in the Activity pane. If necessary, drag the right border of the Activity pane to the right to expand the pane so the entire converted macro name is visible and then double-click the macro name to open the event procedure in its class module window.
3. Read the VBA code and then click the Close button.

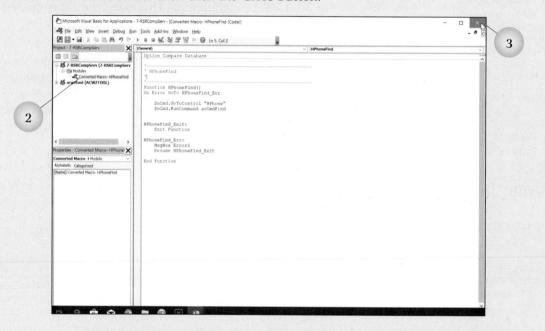

4. Close the HPhoneFind Macro Builder window.

Activity 2 Create a Navigation Form   2 Parts

You will create a Navigation form to be used as a main menu for the RSR Computer Services database.

Creating a
Navigation Form

Creating a Navigation Form

Database files are often accessed by multiple users who use them for specific purposes, such as updating customer records and entering details related to completed work orders. These individuals may not know much about database applications and may want an easy method for completing data entry or maintenance tasks. Users can open the forms and reports needed to update, view, and print data using a Navigation form as a menu. A Navigation form has tabs along the top, left, or right. It can be set to display automatically when the database file is opened so users do not need to choose which objects to open from the Navigation pane.

Quick Steps

Create Navigation Form
1. Click Create tab.
2. Click Navigation button.
3. Click form style.
4. Drag form or report name to *[Add New]* in Navigation Form window.
5. Repeat Step 4 as needed.
6. Click Save.
7. Type form name.
8. Press Enter key.

 Navigation

To create a Navigation form, click the Create tab and then click the Navigation button in the Forms group. At the Navigation button drop-down list, choose the type of form to be created by selecting the option in the drop-down list that positions the tabs where they are to appear. A Navigation Form window opens with a title and a tab bar. Access displays *[Add New]* in the first tab in the form. To add a form or report to a Navigation form, drag it from the Navigation pane and drop it into the to *[Add New]* tab in the Navigation Form window. Continue dragging and dropping form and/or report names from the Navigation pane into the *[Add New]* tab in the Navigation Form window in the order they are to appear. Figure 7.5 illustrates the Navigation form to be created in Activities 2a and 2b.

Figure 7.5 Navigation Form for Activities 2a and 2b

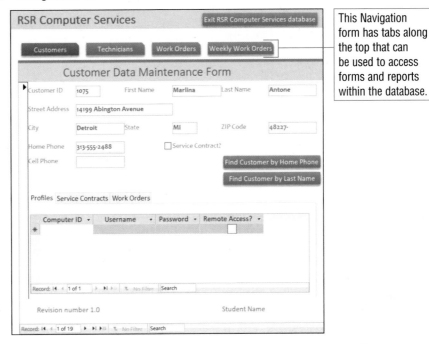

This Navigation form has tabs along the top that can be used to access forms and reports within the database.

Activity 2a Creating a Navigation Form

Part 1 of 2

1. With **7-RSRCompServ** open, create a Navigation form with tabs along the top for accessing forms and reports by completing the following steps:
   a. Click the Create tab.
   b. Click the Navigation button in the Forms group.
   c. Click *Horizontal Tabs* at the drop-down list. Access opens a Navigation Form window with a Field List task pane at the right side of the work area. The horizontal tab at the top of the form is selected and displays *[Add New]*.

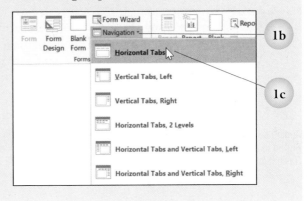

d. Position the mouse pointer on the *CustMaintenance* form name in the Navigation pane, click and hold down the left mouse button, drag the form name to *[Add New]* in the Navigation form, and then release the mouse button. Access adds the CustMaintenance form to the first tab in the form and displays a new tab with *[Add New]* right of the CustMaintenance tab.

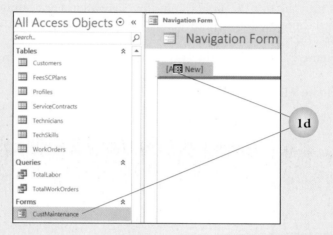

e. Drag the *TechMaintenance* form name from the Navigation pane to the second tab (which displays *[Add New]*).

f. Drag the *WorkOrders* report name from the Navigation pane to the third tab (which displays *[Add New]*).

g. Drag the *WorkOrdersbyWeek* report name from the Navigation pane to the fourth tab (which displays *[Add New]*).

2. Close the Field List task pane at the right side of the work area.

3. Click the Save button on the Quick Access Toolbar. At the Save As dialog box, type MainMenu in the *Form Name* text box and then click OK.

4. Switch to Form view.

5. Click each tab along the top of the Navigation form to view each form or report in the work area. Leave the MainMenu form open for the next activity.

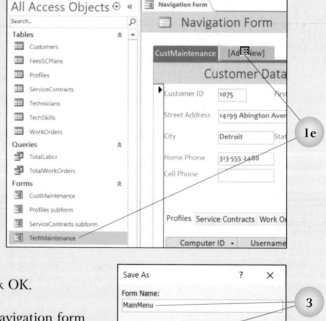

Tutorial

Adding a Command
Button to a
Navigation Form

A Navigation form can be edited in Layout view or Form view using all the tools and techniques discussed in Chapter 4 for working with forms. Consider changing the title, adding a logo and/or a command button, and renaming tabs to customize the Navigation form.

1. With **7-RSRCompServ** open and the MainMenu form displayed in the work area, add a command button to the Navigation form by completing the following steps:
   a. Switch to Design view.
   b. Resize the title control object in the *Form Header* section to align with the right edge of the object at approximately the 2.5-inch mark on the horizontal ruler.
   c. Click the Button button in the Controls group on the Form Design Tools Design tab and then drag to create a button in the Form Header section the approximate height and width shown below.

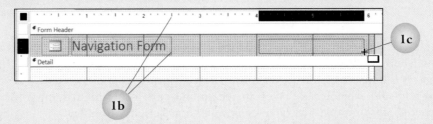

   d. At the first Command Button Wizard dialog box, select *Miscellaneous* in the *Categories* list box, select *Run Macro* in the *Actions* list box, and then click the Next button.
   e. With *ExitRSRdb* already selected in the *Macros* list box at the second Command Button Wizard dialog box, click the Next button.
   f. At the third Command Button Wizard dialog box, click *Text*, select the current entry in the *Text* text box, type Exit RSR Computer Services database, and then click the Next button.
   g. Type Exitdb at the last Command Button Wizard dialog box and then click the Finish button.
   h. With the new command button selected, apply the Intense Effect - Blue, Accent 1 quick style (second column, last row in the *Theme Styles* gallery).
2. Click to select the logo control object left of the title in the *Form Header* section and then press the Delete key.
3. Change the text in the Title control object in the *Form Header* section to *RSR Computer Services* and apply Bold formatting.

4. Relabel the tabs along the top of the Navigation form by completing the following steps:
   a. Click to select the *CustMaintenance* tab and then click the Property Sheet button in the Tools group on the Form Design Tools Design tab.
   b. Click the Format tab in the Property Sheet task pane, select the current text in the *Caption* property box, and then type Customers.
   c. Click the *TechMaintenance* tab, select the current text in the *Caption* property box in the Property Sheet task pane, and then type Technicians.
   d. Click the *WorkOrders* tab, click in the *Caption* property box, and then insert a space between *Work* and *Orders* so the tab name displays as *Work Orders*.
   e. Click the *WorkOrdersbyWeek* tab, select the current text in the *Caption* property box, and then type Weekly Work Orders.
   f. Close the Property Sheet task pane.

5. Select the four named tabs plus the [Add New] tab and then make the following formatting changes:
   a. Change the shape to Rectangle: Single Corner Snipped.
   b. Apply the Intense Effect - Blue, Accent 1 quick style.

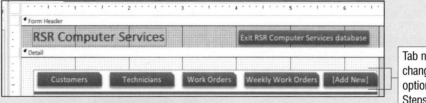

Tab names were changed and formatting options were applied in Steps 4 and 5.

6. Switch to Form view. Insert a screenshot of the database window that shows the form in a new Microsoft Word document using either the Screenshot button in the Illustrations group on the Insert tab or the Windows key + Shift + S and the Paste feature. Type your name a few lines below the screen image and add any other identifying information as instructed (for example, the chapter number and activity number).
7. Save the Microsoft Word document with the name 7-MainMenu, close it, and then exit Word.
8. Save and close the revised form.

 Check Your Work

---

Activity 3 Configure Database Options   7 Parts

You will configure database options for the active database and configure error-checking options for all databases. You will also customize the ribbon and Quick Access Toolbar.

 Tutorial

Customizing the Access Environment

Quick Steps

Set Startup Form
1. Click File tab.
2. Click *Options*.
3. Click *Current Database* in left pane.
4. Click arrow next to *Display Form*.
5. Click form.
6. Click OK.

Specify Application Title
1. Click File tab.
2. Click *Options*.
3. Click *Current Database* in left pane.
4. Click in *Application Title* text box.
5. Type title.
6. Click OK.

Customizing the Access Environment

Click the File tab and then click *Options* to open the Access Options dialog box, which can be used to customize the Access environment. Options can be specified for all databases or only the current database. The behaviors of certain keys and the default margins for printing can also be defined. Choose to set a form to display automatically whenever the database file is opened or to hide the Navigation pane for the current database. For example, if a Navigation form has been created that provides access only to a certain group of objects, hide the Navigation pane for the current database. A database can be set to open by default in shared use or exclusive use. *Exclusive use* means the file is restricted to one individual user.

Figure 7.6 displays the Access Options dialog box with *Current Database* selected in the left pane. With options for the current database displayed, define a startup form to open automatically when the database is opened. Activity 3a demonstrates how to configure the current database to display the MainMenu form when the database is opened and to specify an application title to display in the Title bar when the database is open. The Navigation pane is customized in Activity 3b.

Figure 7.6 Access Options Dialog Box with *Current Database* Selected

Customize the active database using options in this section.

Click this button to open the Navigation Options dialog box, where the Navigation pane can be customized.

Remove this check mark to hide the Navigation pane in the current database.

Activity 3a Specifying a Startup Form and Application Title

Part 1 of 7

1. With **7-RSRCompServ** open, specify an application title to display in the Title bar and choose a form to display automatically when the database file is opened by completing the following steps:
   a. Click the File tab.
   b. Click *Options* to open the Access Options dialog box.
   c. Click *Current Database* in the left pane.
   d. Click in the *Application Title* text box in the *Application Options* section and then type RSR Computer Services Database.
   e. Click the arrow next to the *Display Form* option box (which displays [*none*]) and then click *MainMenu* at the drop-down list.
2. Click OK.
3. Click OK at the Microsoft Access message box indicating that the database must be closed and reopened for the options to take effect.

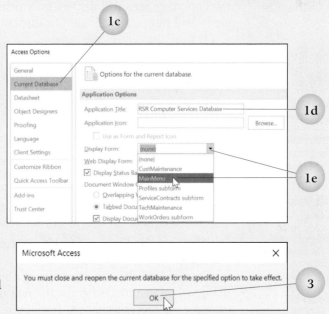

4. Close **7-RSRCompServ**.
5. Reopen **7-RSRCompServ**. The MainMenu form displays automatically in the work area and the Title bar displays the application title *RSR Computer Services Database*.

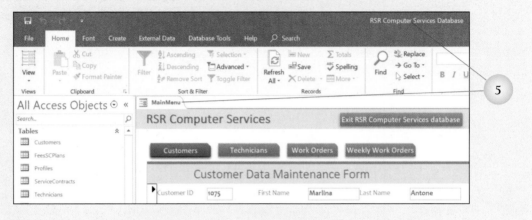

Tutorial

Limiting Ribbon Tabs and Menus in a Database

Limiting Ribbon Tabs and Menus in a Database

In addition to limiting users' access to objects, access to options in the ribbon and menus can be limited. Restricting users from seeing the full ribbon and shortcut menus will prevent accidental changes from being made if someone switches views and edits or deletes objects without knowing the full impact of these changes.

To limit users' access, display the Access Options dialog box and select *Current Database* in the left pane. Scroll down the dialog box to the *Ribbon and Toolbar Options* section. Click to remove the check marks from the *Allow Full Menus* and *Allow Default Shortcut Menus* check boxes and then click OK. When the database is closed and reopened, only the Home tab will display in the ribbon. The File tab backstage area will display only print options. Users will not be able to switch views and right-clicking will not display a shortcut menu.

If full access to the ribbon and menus is needed, bypass the startup options. Do this by pressing and holding down the Shift key while double-clicking the file name in the File Open backstage area when opening the database.

Tutorial

Customizing the Navigation Pane

Customizing the Navigation Pane

When using a startup form in a database, consider hiding the Navigation pane to prevent users from accidentally making changes to other objects. To hide the Navigation pane, open the Access Options dialog box, click *Current Database* in the left pane, and then click to remove the check mark from the *Display Navigation Pane* check box in the *Navigation* section. When the Navigation pane is hidden, use the F11 function key to unhide it.

Click the Navigation Options button in the *Navigation* section to open the Navigation Options dialog box, shown in Figure 7.7. At this dialog box, choose to hide individual objects or groups of objects, set display options for the pane, and define whether objects can be opened using a single or double mouse click. For example, to prevent changes from being made to the table design, hide the Tables group.

Quick Steps

Hide Navigation Pane
1. Click File tab.
2. Click *Options*.
3. Click *Current Database* in left pane.
4. Clear *Display Navigation Pane* check box.
5. Click OK two times.

Customize Navigation Pane

1. Click File tab.
2. Click *Options*.
3. Click *Current Database* in left pane.
4. Click Navigation Options button.
5. Select options.
6. Click OK two times.

Figure 7.7 Navigation Options Dialog Box

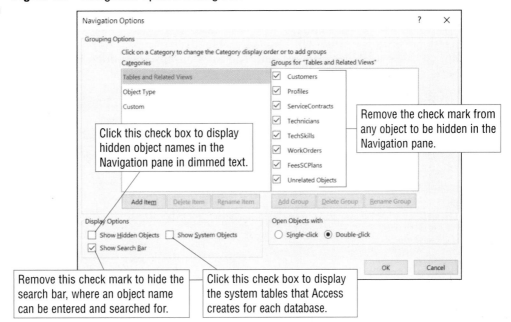

Click this check box to display hidden object names in the Navigation pane in dimmed text.

Remove the check mark from any object to be hidden in the Navigation pane.

Remove this check mark to hide the search bar, where an object name can be entered and searched for.

Click this check box to display the system tables that Access creates for each database.

Activity 3b Customizing the Navigation Pane

Part 2 of 7

1. With **7-RSRCompServ** open, customize the Navigation pane to hide all the table, macro, and module objects by completing the following steps:

   a. Click the File tab and then click *Options*.

   b. If necessary, click *Current Database* in the left pane. Click the Navigation Options button in the *Navigation* section. (You may need to scroll down to view the *Navigation* section.)

   c. Click *Object Type* in the *Categories* list box.

   d. Click the *Tables* check box in the *Groups for "Object Type"* list box to remove the check mark.

   e. Remove the check mark from the *Macros* check box.

   f. Remove the check mark from the *Modules* check box.

   g. Click OK to close the Navigation Options dialog box.

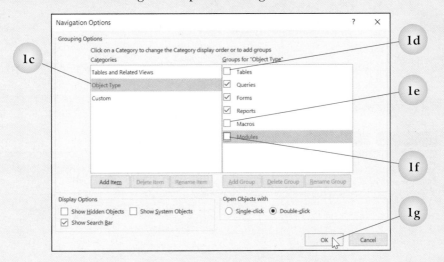

h. Click OK to close the Access Options dialog box.

i. Click OK at the message box stating that the current database must be closed and reopened for the specified option to take effect.

2. Notice that the Tables, Macros, and Modules groups are hidden in the Navigation pane.

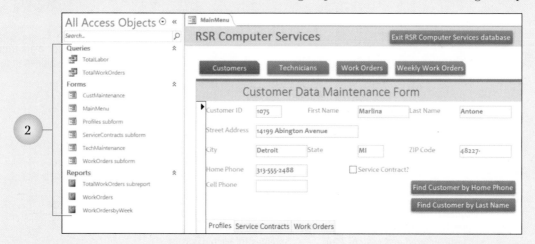

3. After reviewing the customized Navigation pane, you have decided that the database will be more secure if the entire pane is hidden when the database is opened. To do this, complete the following steps:

a. Click the File tab and then click *Options*.

b. With *Current Database* already selected in the left pane, click the *Display Navigation Pane* check box in the *Navigation* section to remove the check mark.

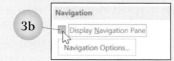

c. Click OK.

d. Click OK at the message box indicating that the database must be closed and reopened for the option to take effect.

4. Close **7-RSRCompServ**.

5. Reopen **7-RSRCompServ**. The Navigation pane is hidden and the MainMenu form is open in the work area.

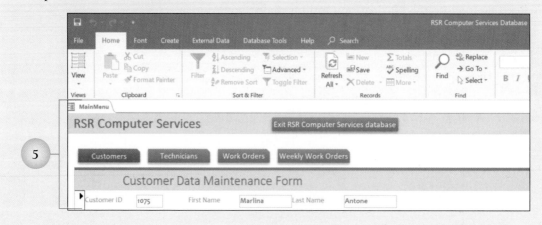

Configuring Error-Checking Options

Recall from Chapter 5 that a green diagonal triangle displays in the Report Selector button when the report is wider than the page allows. Clicking the error-checking options button provides access to tools that can fix the report automatically. A green triangle also appears in a new label control object that is added to a report without being associated with another control object. Access flags the label as an error because it is not associated with another control object. Unchecking the *Check for unassociated label and control option* stops Access from flagging the control object.

By default, Access has error checking turned on and all error-checking options active. To configure error-checking options in Access, open the Access Options dialog box and select *Object Designers* in the left pane. Scroll down the right pane to locate the *Error checking in form and report design view* section, shown in Figure 7.8. Remove the check marks in the check boxes for those options to be disabled and then click OK. Table 7.1 provides a description of each option.

Figure 7.8 Error-Checking Options in Access

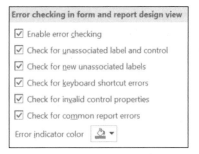

Table 7.1 Error-Checking Options

Error-Checking Option	Description
Enable error checking	Error checking can be turned on or off in forms and reports. An error is indicated by a green triangle in the upper left corner of a control object.
Check for unassociated label and control	A selected label and text box control object are checked to make sure the two control objects are associated. A Trace Error button appears if Access detects an error.
Check for new unassociated labels	Each new label control object is checked for association with a text box control object.
Check for keyboard shortcut errors	Duplicate keyboard shortcuts and invalid shortcuts are flagged.
Check for invalid control properties	Invalid properties, formula expressions, and field names are flagged.
Check for common report errors	Reports are checked for errors, such as invalid sort orders and reports that are wider than the selected paper size.
Error indicator color	A green triangle indicates an error in a control. Click the Color Picker button to change to a different color.

1. With **7-RSRCompServ** open, assume that label control objects that contain explanatory text for users are frequently added to forms and reports. Customize the error-checking options to prevent Access from flagging these independent label control objects as errors. To do this, complete the following steps:

a. Click the File tab and then click *Options*.

b. Click *Object Designers* in the left pane.

c. Scroll down the right pane to the *Error checking in form and report design view* section.

d. Click the *Check for new unassociated labels* check box to remove the check mark.

e. Click OK.

f. Click OK.

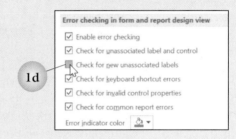

Customizing the Ribbon

Customizing the Ribbon

Hint To save mouse clicks, consider creating a custom tab that contains the buttons used on a regular basis.

Customize the ribbon by creating a new tab. Within the new tab, add groups and then add buttons within the groups. To customize the ribbon, click the File tab and then click *Options*. At the Access Options dialog box, click *Customize Ribbon* in the left pane. Options for customizing the ribbon are shown in Figure 7.9.

Figure 7.9 Access Options Dialog Box with *Customize Ribbon* Selected

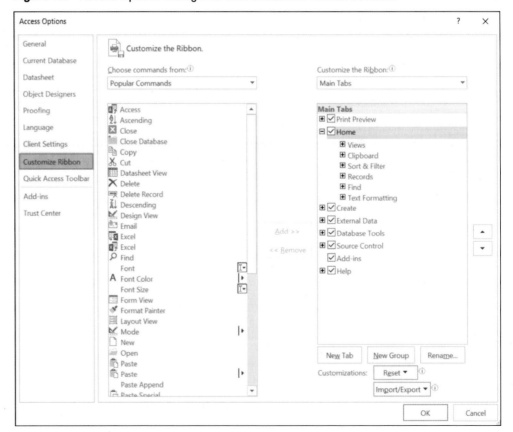

The commands shown in the left list box in Figure 7.9 are dependent on the current option selected in the *Choose commands from* option box above it. Click the arrow right of the current option (*Popular Commands*) to select from a variety of command lists, such as *Commands Not in the Ribbon* and *All Commands*. The tabs shown in the right list box are dependent on the current option selected in the *Customize the Ribbon* option box. Click the arrow right of the current option (*Main Tabs*) to select *All Tabs*, *Main Tabs*, or *Tool Tabs*.

Create a new group in an existing tab or create a new tab along with a new group within the tab. Add buttons that are regularly used to either of the new groups.

Creating a New Tab

Quick Steps

Create New Tab and Group

1. Click File tab.
2. Click *Options*.
3. Click *Customize Ribbon* in left pane.
4. Click tab name to precede new tab.
5. Click New Tab button.

Add a New Group to an Existing Tab

1. Click File tab.
2. Click *Options*.
3. Click *Customize Ribbon* in left pane.
4. Click tab name with which new group is associated.
5. Click New Group button.

To create a new tab, click the tab name in the *Main Tabs* list box that the new tab is to follow and then click the New Tab button below the list box. This inserts a new tab in the list box along with a new group below the new tab, as shown in Figure 7.10. If the wrong tab name was selected before clicking the New Tab button, the new tab can be moved up or down in the list box. To do this, click *New Tab (Custom)* and then click the Move Up or Move Down buttons that display at the right side of the dialog box.

Figure 7.10 New Tab and Group Created in the Customize Ribbon Pane at the Access Options Dialog Box

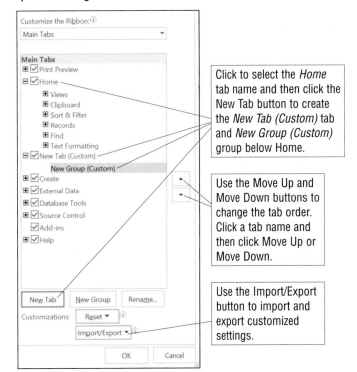

Click to select the *Home* tab name and then click the New Tab button to create the *New Tab (Custom)* tab and *New Group (Custom)* group below Home.

Use the Move Up and Move Down buttons to change the tab order. Click a tab name and then click Move Up or Move Down.

Use the Import/Export button to import and export customized settings.

Exporting Customizations

Your educational institution may already have taken the time to customize the ribbon in Access. To ensure that these customizations can be restored after the ribbon is changed in Activity 3e, the settings will be saved (exported) in Activity 3d. In Activity 3g, the original settings will be reinstalled (imported).

To save the current ribbon settings, click the File tab and then click *Options*. At the Access Options dialog box, click *Customize Ribbon* in the left pane. Click the Import/Export button in the lower right corner of the Access Options dialog box. Click *Export all customizations* to save the file with the custom settings. This file will be used to reinstall the saved settings; it could also be used to install the customized settings on a different computer. To import the saved settings, click the Import/Export button and then click *Import customization file*. Locate the file and reinstall the customized settings.

Renaming a Tab or Group

Rename a tab by clicking the tab name in the *Main Tabs* list box and then clicking the Rename button below the list box. At the Rename dialog box, type the name for the tab and then press the Enter key or click OK. The Rename dialog box can also be displayed by right-clicking the tab name and then clicking *Rename* at the shortcut menu.

Complete similar steps to rename a group. The Rename dialog box for a group name or command name contains a *Symbol* list box as well as a *Display name* text box. However, the symbols are more useful for identifying new buttons than new groups. Type the new name for the group in the *Display name* text box and then press the Enter key or click OK.

Adding Buttons to a Tab Group

Add commands to a tab by clicking the group name within the tab, clicking the desired command in the list box at the left, and then clicking the Add button between the two list boxes. Remove commands in a similar manner but click the Remove button instead.

Activity 3d Exporting Customizations

1. With **7-RSRCompServ** open, save the current ribbon settings to the desktop by completing the following steps:
   a. Click the File tab and then click *Options*.
   b. Click *Customize Ribbon* in the left pane of the Access Options dialog box.
   c. Click the Import/Export button in the bottom right of the Access Options dialog box.
   d. Click *Export all customizations* at the drop-down list.
   e. Click *Desktop* in the *Favorites* list at the left side of the File Save dialog box.
   f. Change the file name to **7-AccessCustomizations** and then click the Save button.
   g. Click OK two times.

1. With **7-RSRCompServ** open, customize the ribbon by adding a new tab and two new groups on the tab by completing the following steps. *Note: The original ribbon settings will be restored in Activity 3g.*

   a. Click the File tab and then click *Options*.

   b. If necessary click *Customize Ribbon* in the left pane of the Access Options dialog box.

   c. Click *Home* in the *Main Tabs* list box.

   d. Click the New Tab button below the list box. (This inserts a new tab below *Home* and a new group below the new tab.)

   e. With *New Group (Custom)* selected below *New Tab (Custom)*, click the New Group button below the list box. (This inserts another new group below the new tab.)

2. Rename the tab and groups by completing the following steps:

   a. Click to select *New Tab (Custom)* in the *Main Tabs* list box.

   b. Click the Rename button below the list box.

   c. At the Rename dialog box, type your first and last names and then click OK.

   d. Click to select the first *New Group (Custom)* group name below the new tab.

   e. Click the Rename button.

   f. At the Rename dialog box, type Views in the *Display name* text box and then click OK. (Notice that the Rename dialog box for a group or button displays symbols in addition to the *Display name* text box. You will apply a symbol to a button in a later step).

   g. Right-click the *New Group (Custom)* group name below *Views (Custom)* and then click *Rename* at the shortcut menu.

   h. Type Records in the *Display name* text box at the Rename dialog box and then click OK.

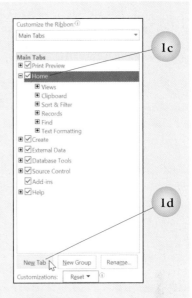

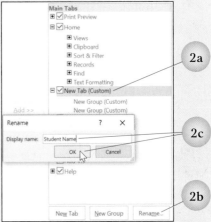

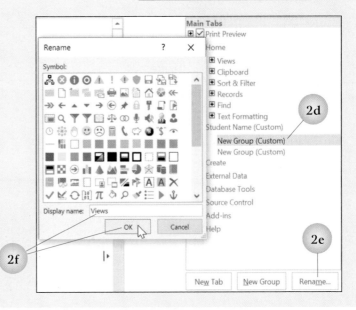

3. Add buttons to the Views (Custom) group by completing the following steps:
   a. Click to select *Views (Custom)* in the *Main Tabs* list box.
   b. With *Popular Commands* selected in the *Choose commands from* option box, click *Form View* in the list box and then click the Add button between the two list boxes. This inserts the command below the *Views (Custom)* group name.

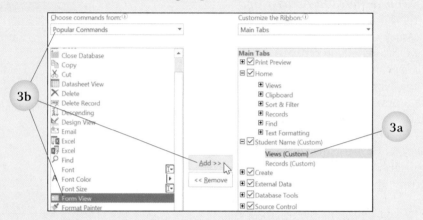

   c. Scroll down the *Popular Commands* list box; click the second *Print Preview* option, which displays the ScreenTip *Popular Commands | Print Preview (FilePrintPreview)*, and then click the Add button. ***Note: The commands are organized in alphabetical order.***
   d. Click *Report View* in the *Popular Commands* list box and then click the Add button.
   e. Scroll to the bottom of the *Popular Commands* list box, click *View*, and then click the Add button.

4. Add buttons to the Records (Custom) group by completing the following steps:
   a. Click to select *Records (Custom)* in the *Main Tabs* list box.
   b. Click the *Choose commands from* option box arrow (which displays *Popular Commands*) and then click *All Commands* at the drop-down list.
   c. Scroll down the *All Commands* list box, click *Ascending*, and then click the Add button.
   d. Scroll down the *All Commands* list box (the list displays alphabetically), click *Delete Record*, and then click the Add button.
   e. Scroll down the *All Commands* list box, click *Find*, and then click the Add button.
   f. Scroll down the *All Commands* list box, click the first *New* option (which displays the ScreenTip *Home Tab | Records | New (GoToNewRecord)*), and then click the Add button.
   g. Scroll down the *All Commands* list box, click *Spelling*, and then click the Add button.

5. Change the symbol for the Spelling button by completing the following steps:
   a. Right-click *Spelling* in the *Records (Custom)* group in the *Main Tabs* list box.
   b. Click *Rename* at the shortcut menu.
   c. At the Rename dialog box click the purple book icon in the *Symbol* list box (eleventh column eighth row, position may vary).
   d. Click OK.

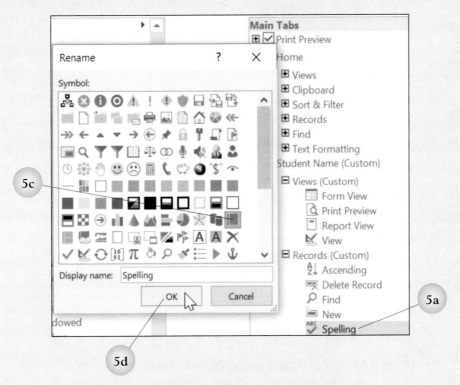

6. Click OK to close the Access Options dialog box. If a message displays stating that the database must be closed and reopened for the option to take effect click OK.
7. Use buttons in the custom tab to change views and spell check a form by completing the following steps:
   a. Click the Work Orders tab at the top of the MainMenu form.
   b. Click the custom tab with your name and then click the Print Preview button in the Views group.
   c. Click the View button in the Views group.
   d. Click the Customers tab in the Main Menu form, click in the *Last Name* field, click the custom tab with your name, and then click the Spelling button.
   e. Click the Cancel button at the Spelling dialog box.
   f. Click the Customers tab at the top of the MainMenu form.
8. Insert a screenshot of the database window that shows the custom tab in a new Microsoft Word document using either the Screenshot button in the Illustrations group on the Insert tab or the Windows key + Shift + S and the Paste feature. Type your name a few lines below the screen image and add any other identifying information as instructed (for example, the chapter number and activity number).
9. Save the Microsoft Word document and name it **7-CustomRibbon**.
10. Print **7-CustomRibbon** and then exit Word.

Customizing the Quick Access Toolbar

Customize Quick Access Toolbar

Click the Customize Quick Access Toolbar button at the right side of the Quick Access Toolbar to open the Customize Quick Access Toolbar drop-down list, as shown in Figure 7.11. Click *More Commands* at the drop-down list to open the Access Options dialog box with *Quick Access Toolbar* selected in the left pane, as shown in Figure 7.12. Change the list of commands that display in the left list box by clicking *Choose commands from* option box arrow and then clicking the appropriate category. Scroll down the list box to locate the command and then double-click the command name to add it to the Quick Access Toolbar.

Quick Steps

Add Button to Quick Access Toolbar
1. Click Customize Quick Access Toolbar button.
2. Click button.
OR
1. Click Customize Quick Access Toolbar button.
2. Click *More Commands*.
3. Click *Choose commands from* option box arrow.
4. Click category.
5. Click command in commands list box.
6. Click Add.
7. Click OK.

Figure 7.11 Customize Quick Access Toolbar Drop-Down List

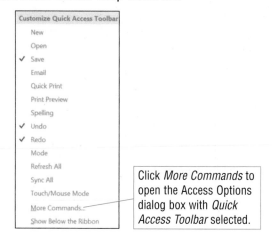

Click *More Commands* to open the Access Options dialog box with *Quick Access Toolbar* selected.

Figure 7.12 Access Options Dialog Box with *Quick Access Toolbar* Selected

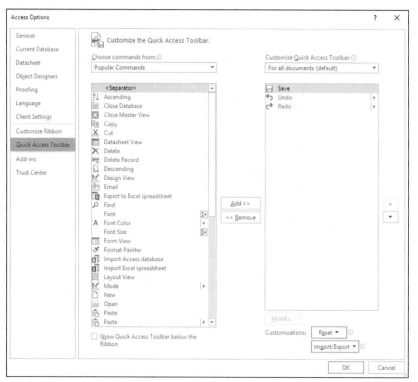

Delete a button from the Quick Access Toolbar by clicking the Customize Quick Access Toolbar button and then clicking the command at the drop-down list. If the command does not appear in the drop-down list, click the *More Commands* option. At the Access Options dialog box, click the command to be removed in the right list box and then click the Remove button.

Activity 3f Adding Commands to the Quick Access Toolbar

Part 6 of 7

1. With **7-RSRCompServ** open, add the Print Preview and Close Window commands to the Quick Access Toolbar by completing the following steps. *Note: The original settings for the Quick Access Toolbar will be restored in Activity 3g.*

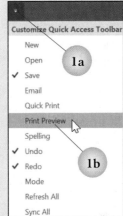

   a. Click the Customize Quick Access toolbar button at the right side of the Quick Access Toolbar.
   b. Click *Print Preview* at the drop-down list. The Print Preview button is added to the end of the Quick Access Toobar. ***Note: Skip to Step 1d if the Print Preview button already displays on your Quick Access Toolbar.***
   c. Click the Customize Quick Access Toobar button.
   d. Click *More Commands* at the drop-down list.
   e. At the Access Options dialog box with *Quick Access Toolbar* selected in the left pane, click the *Choose commands from* option box arrow and then click *Commands Not in the Ribbon.*

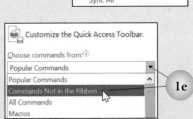

   f. Double-click *Close Window* in the *Commands Not in the Ribbon* list box. ***Note: The commands are organized in alphabetical order.***
   g. Click OK. If a message displays stating that the database must be closed and reopened for the option to take effect, click OK. The Close Window button is added right of the Save button.
2. Click the WorkOrders tab in the MainMenu Navigation form.
3. Click the PrintPreview button on the Quick Access Toolbar to open the WorkOrders report in Print Preview. Click Close Print Preview.
4. Click the Close Window button on the Quick Access Toolbar.

Quick Steps
Reset Ribbon and Quick Access Toolbar
1. Click the File tab.
2. Click *Options.*
3. Click *Customize Ribbon.*
4. Click Reset button.
5. Click *Reset all customizations.*
6. Click Yes.
7. Click OK.

Resetting the Ribbon and the Quick Access Toolbar

Restore the original ribbon and the Quick Access Toolbar by clicking the Reset button below the *Main Tabs* list box in the Access Options dialog box with *Customize Ribbon* selected in the left pane. Clicking the Reset button displays these two options: *Reset only selected Ribbon tab* and *Reset all customizations*. Click *Reset all customizations* to restore the ribbon and the Quick Access Toolbar to its original settings and then click Yes at the Microsoft Office message box that displays the message *Delete all Ribbon and Quick Access Toolbar customizations for this program?*

To restore the ribbon and the Quick Access Toolbar to your institution's customized settings, import the settings that were exported in Activity 3d. Click the Import/Export button in the lower right corner of the Access Options dialog box and then click *Import customization file* at the drop-down list. Locate the file and reinstall the customized settings.

Activity 3g Restoring the Ribbon and the Quick Access Toolbar   Part 7 of 7

1. Import the customization file saved in Activity 3d to reset the ribbon and the Quick Access Toolbar to your institution's original settings by completing the following steps:
   a. Click the File tab and then click *Options* to open the Access Options dialog box.
   b. If necessary, click *Customize Ribbon* in the left pane.
   c. Click the Import/Export button in the bottom right corner of the dialog box.
   d. Click *Import customization file* at the drop-down list.
   e. Click *Desktop* in the *Favorites* list at the left of the File Open dialog box.
   f. Click **7-AccessCustomizations.exportedUI**.
   g. Click the Open button.
   h. Click Yes at the Microsoft Office message box that asks about replacing all existing ribbon and Quick Access Toolbar customizations.

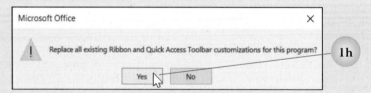

   i. Click OK to close the Access Options dialog box.
   j. Click OK at the message that displays saying that the database must be closed and reopened for the option to take effect.
2. If **7-RSRCompServ** is open, close it.

Chapter Summary

- A macro can be used to automate actions within a database, such as opening a form, query, or report.

- To create a macro, click the Create tab and then click the Macro button in the Macros & Code group, which opens a Macro Builder window. Add a macro action using the *Add New Action* option box or Action Catalog task pane.

- Action arguments are the parameters for the action, such as the object name, mode in which the object opens, and other restrictions placed on the action.

- The arguments displayed vary depending on the active action.

- Most macros can be run by double-clicking the macro name in the Navigation pane, right-clicking the macro name, or clicking the Run button in the Macro Builder window.

- A macro can also be created by dragging and dropping an object name from the Navigation pane into the *Add New Action* option box in a Macro task pane.

- Edit a macro by right-clicking the macro name in the Navigation pane and clicking *Design View* at the shortcut menu.

- A macro can be assigned to a button in a form to allow running the macro with a single mouse click.

- Use the Button button to create a command button in a form.

- Use the Command Button Wizard to create an embedded macro that tells Access which macro to run when the button is clicked. View the macro by opening the command button's Property Sheet task pane, clicking the Event tab, and then clicking the Build button in the *On Click* property box.

- Convert a macro to Visual Basic for Applications (VBA) code by opening it in the Macro Builder window and then clicking the Convert Macros to Visual Basic button in the Tools group on the Macro Tools Design tab.

- A Navigation form has tabs along the top, left, or right and is used as a menu. Users can open the forms and reports in the database by clicking a tab, rather than using the Navigation pane.

- To create a Navigation form, click the Create tab and then click the Navigation button in the Forms group. Choose the form type at the drop-down list and then drag and drop form and/or report names from the Navigation pane into *[Add New]* in the tab bar.

- A form can be set to display automatically when the database is opened. Select the startup form at the *Display Form* option box in the Access Options dialog box with *Current Database* selected in the left pane.

- Change the title that appears in the Title bar for the active database by typing text in the *Application Title* text box at the Access Options dialog box with *Current Database* selected in the left pane.

- Set options for the Navigation pane, such as hiding individual objects and groups of objects, at the Navigation Options dialog box.

- Hide the Navigation pane by removing the check mark from the *Display Navigation Pane* check box at the Access Options dialog box with *Current Database* selected in the left pane.

- Change the default error-checking options at the Access Options dialog box with *Object Designers* selected in the left pane.

- Customize the ribbon by creating a new tab, creating a new group on the new tab, and then adding buttons within the new group.
- To customize the ribbon, open the Access Options dialog box and click *Customize Ribbon* in the left pane.
- To save the current ribbon and Quick Access Toolbar settings, open the Access Options dialog box, click *Customize Ribbon* in the left pane, click the Import/Export button, and then click *Export all customizations*. This file can then be used to reinstall the customizations.
- Create a new tab by clicking the tab name that will precede the new tab in the *Main Tabs* list box and then clicking the New Tab button. A new group is automatically added with the new tab.
- Rename a tab by clicking the tab name in the *Main Tabs* list box, clicking the Rename button, typing a new name, and then pressing the Enter key or clicking OK. Rename a group using a similar process.
- Add buttons to a tab by clicking the group name within the tab, selecting the desired command in the commands list box, and then clicking the Add button between the two list boxes.
- Add buttons to or delete buttons from the Quick Access Toolbar using the Customize Quick Access Toolbar button. To locate a feature to add, click the *More Commands* at the drop-down list to open the Access Options dialog box with *Quick Access Toolbar* selected.
- Restore the ribbon to the default settings by clicking the Reset button below the *Main Tabs* list box in the Access Options dialog box with *Customize Ribbon* selected and then clicking *Reset all customizations* at the drop-down list.

Commands Review

FEATURE	RIBBON TAB, GROUP	BUTTON
convert macros to Visual Basic for Applications	Macro Tools Design, Tools	
create command button	Form Design Tools Design, Controls	
create macro	Create, Macros & Code	
create Navigation form	Create, Forms	
customize Navigation pane	File, Options	
run macro	Macro Tools Design, Tools	

Access®

Integrating Access Data

Performance Objectives

Upon successful completion of Chapter 8, you will be able to:

1 Create and restore a backup database file

2 Create an ACCDE database file

3 View Trust Center settings

4 Import and merge data from another Access database

5 Link to a table in another Access database

6 Determine when to import or link to external source data

7 Reset or refresh a link using the Linked Table Manager

8 Import data from a text file

9 Save and repeat import specifications

10 Export Access data to a text file

11 Save and repeat export specifications

Integrating Access data with other applications in the Microsoft Office suite is easily accomplished with buttons on the External Data tab. These buttons allow you to export data to and import data from Word and Excel files. Data can be exchanged between the Microsoft programs with the formatting and data structure intact. In some cases, however, you may need to exchange data between Access and a non-Microsoft program. In this chapter, you will learn how to create and restore a backup database file, how to integrate data between individual Access database files, and how to import and export data in a text file format that is recognized by nearly all applications. You will also learn how to prevent changes from being made to the design of objects.

 Data Files

Before beginning chapter work, copy the AL2C8 folder to your storage medium and then make AL2C8 the active folder.

The online course includes additional training and assessment resources.

Activity 1 Maintain and Secure a Database   3 Parts

You will back up and restore a database and secure a database by making an ACCDE file. You will also explore the default settings in the Trust Center.

 Tutorial

Review: Backing Up a Database

Quick Steps

Back Up Database
1. Click File.
2. Click *Save As*.
3. Click *Back Up Database*.
4. Click Save As button.
5. Navigate to appropriate folder.
6. Click Save button.

Creating and Restoring a Backup File

Before making any major changes to a database, such as running action queries or modifying table structures, it is a good idea to back it up. Certain changes made to a database cannot be reversed. For example, if a delete query is run to remove the work orders for the month of September, the work orders cannot be reinstated. If field sizes are changed and made too small, the data truncated cannot be restored if the revised table has been saved. The data will be lost unless a backup exists.

Consider backing up your database on a regular basis to minimize any data loss that may occur due to system failures or design mistakes. Once a backup database has been created, use it to restore the entire database or to import a specific object. Activity 1 will demonstrate how to restore the entire database and Activity 2 will demonstrate how to import objects.

To create a backup of a database, click the File tab and then click the *Save As* option. Click *Back Up Database* in the *Advanced* section of the Save As backstage area and then click the Save As button. (Access closes any open objects.) Navigate to or create a folder to store backup databases and then click the Save button. Notice that the current date is added to the end of the file name.

To replace a database with a backup copy, copy the backup database from the backup folder and then paste it in the same folder as the database to be replaced. Delete the database to be replaced and rename the backup database.

Activity 1a Creating and Restoring a Backup File   Part 1 of 3

1. Open **8-RSRCompServ** and enable the content.
2. Create a backup database file by completing the following steps:
   a. Click the File tab and then click the *Save As* option.
   b. Click *Back Up Database* in the *Advanced* section.
   c. Click the Save As button.

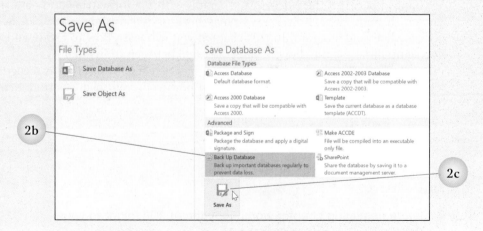

d. The backup file is named **8-RSRCompServ_yyyy-mm-dd**, where *yyyy-mm-dd* represents the year, month, and day the backup was created. (Your date will differ from the one shown below.) Notice that the default location is the AL2C8 folder on your storage medium. Instead of saving the file in this folder, click the New folder button.

e. With *New folder* selected, type DatabaseBackUps and then press the Enter key.

f. Click the Open button to open the folder.

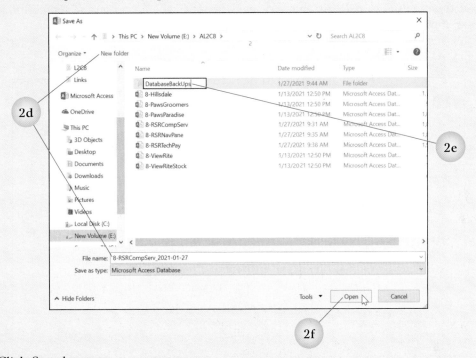

g. Click Save button.

3. With **8-RSRCompServ** open, open the Customers table in Design view, click *City*, and then change the field size from 25 characters to 5 characters. Save the table.

4. Click the Yes button to the Microsoft Access message box stating that some data may be lost.

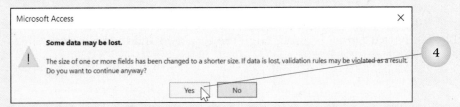

5. Switch to Datasheet view and look at the data in the *City* column. The last two letters were permanently deleted from *Detroit*. Notice that the Undo button on the Quick Access Toolbar is dimmed, indicating that this change cannot be reversed.

6. Close **8-RSRCompServ** and then close Access.

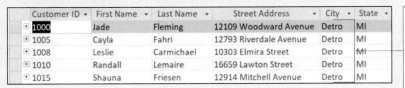

The field size for the *City* field has been changed to 5 characters. The last two letters in *Detroit* have been permanently deleted.

7. Restore the backup copy of the database by completing the following steps:
   a. Click the Start button.
   b. At the Start screen, start typing file explorer. When the File Explorer icon displays below the *Apps* heading, press the Enter key. (Depending on your operating system, these steps may vary.)
   c. Navigate to the DatabaseBackUps folder in the AL2C8 folder on your storage medium and then right-click **8-RSRCompServ_yyyy-mm-dd**, where *yyyy-mm-dd* represents the year, month, and day the backup was created.
   d. Click *Copy* at the shortcut menu.
   e. Click *AL2C8* in the Address bar.

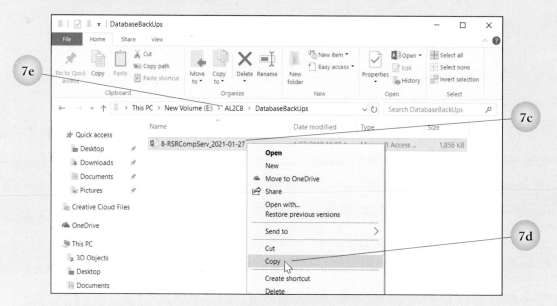

   f. Right-click a blank area of the content pane and then click *Paste* at the shortcut menu.
   g. Right-click **8-RSRCompServ** and then click *Delete* at the shortcut menu.
   h. Right-click **8-RSRCompServ_yyyy-mm-dd** and then click *Rename* at the shortcut menu.
   i. Delete *_yyyy-mm-dd* and then press the Enter key.
8. Open **8-RSRCompServ**, enable the content if necessary, and then open the Customers table. Verify that *Detroit* is displayed in the *City* field and then close the table.
9. Close **8-RSRCompServ**.

 Tutorial

Creating an ACCDE Database File

💡**Hint** ACCDE stands for "Access Database Execute Only." ACCDE files have the file extension .accde instead of .accdb. To see the different file extensions, turn on the display of file extensions at the File Explorer window.

Creating an ACCDE Database File

In Chapter 6, a database was split into two files to create a front-end database and a back-end database. Using this method improved the performance of the database and protected the table objects from being changed by separating them from the queries, forms, and reports. Another method to protect an Access database is to create an ACCDE file. In an ACCDE file, end users are prevented from making changes to the designs of objects. An Access database stored as an ACCDE file is a locked-down version of the database and therefore does not provide access to Design view or Layout view. In addition, if the database contains any Visual Basic for Applications (VBA) code, the code cannot be modified or changed.

Create ACCDE File
1. Open database.
2. Click File tab.
3. Click *Save As*.
4. Click *Make ACCDE*.
5. Click Save As button.
6. Navigate to required drive and/or folder.
7. Type name in *File name* text box.
8. Click Save button.

To save an Access database as an ACCDE file, click the File tab and then click the *Save As* option. Click the *Make ACCDE* option in the *Advanced* section of the Save As backstage area and then click the Save As button to open the Save As dialog box. Navigate to the drive and/or folder in which to save the database, type the file name in the *File name* text box, and then click the Save button. Turn on the display of file extensions at the File Explorer window by clicking the View tab and then clicking to add a check mark in the *File extensions* check box in the Show/hide group. Note that the newly saved database file has the extension .accde. Move the original database with the .accdb extension to a secure location and provide end users with the path to the ACCDE file for daily use.

Activity 1b Creating an ACCDE Database File

Part 2 of 3

1. Open **8-RSRNavPane** and enable the content.
2. Create an ACCDE file by completing the following steps:
   a. Click the File tab and then click the *Save As* option.
   b. Click *Make ACCDE* in the *Advanced* section of the Save As backstage area.
   c. Click the Save As button.

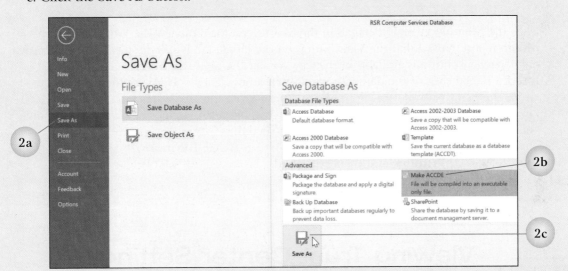

   d. At the Save As dialog box with the default location the AL2C8 folder on your storage medium add the text *NoChanges* to the end of the file name so that it reads **8-RSRNavPaneNoChanges** in the *File name* text box. Click the Save button.

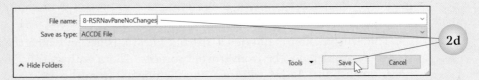

3. Close **8-RSRNavPane**.
4. Open **8-RSRNavPaneNoChanges**.

5. At the Microsoft Access Security Notice dialog box stating that the file might contain unsafe content, click the Open button.
6. With Customers the active tab in the MainMenu form and the Customer Data Maintenance Form open in Form view, click the Home tab if necessary. Notice that the View button in the Views group is dimmed.

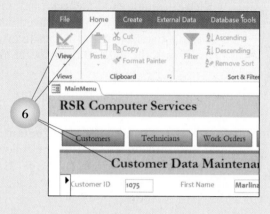

7. Click the Monthly Work Orders tab in the MainMenu form to view the Work Orders by Month report. Notice that the View button on the Home tab is still dimmed. Also notice that only one view button is available in the view area at the right side of the Status bar.
8. Insert a screen shot of the database window in a new Microsoft Word document using either the Screenshot button in the Illustrations group on the Insert tab or the Windows key + Shift + S and the Paste feature. Type your name a few lines below the screen image and add other identifying information as instructed.
9. Save the Microsoft Word document and name it **8-ACCDEWindow**.
10. Print **8-ACCDEWindow** and then exit Word.
11. Close **8-RSRNavPane**.

Viewing Trust Center Settings

Viewing Trust Center Settings

In Access, the Trust Center is set to block unsafe content when a database is opened. Throughout these chapters, the security warning that appears in the message bar when a database is opened has been closed by clicking the Enable Content button. Access provides the Trust Center to allow viewing and/or modifying the security options that are in place to protect your computer from malicious content.

The Trust Center maintains a Trusted Locations list, in which the content stored is considered to be from trusted sources. Add a path to the Trusted Locations list and Access will treat any file opened from the drive and folder as safe. A database opened from a trusted location does not display the security warning in the message bar and does not have content blocked.

Before the macros are enabled for a database, the Trust Center checks for a valid and current digital signature signed by an entity that is stored in the Trusted Publishers list. The Trusted Publishers list is maintained by the user on the computer being used. A trusted publisher is added to the list when the content is enabled from an authenticated source and the *Trust all content from this publisher* option has been clicked. Depending on the active macro security setting, if the

Quick Steps

View Trust Center Options
1. Click File tab.
2. Click *Options*.
3. Click *Trust Center* in left pane.
4. Click Trust Center Settings button.
5. Click desired Trust Center category in left pane.
6. View and/or modify options.
7. Click OK two times.

Trust Center cannot match the digital signature information with an entity in the Trusted Publishers list or if the macro does not contain a digital signature, the security warning displays in the message bar.

Table 8.1 describes the four options available for macro security. The default macro security option is *Disable all macros with notification*. In some cases, it may be decided to change the default macro security setting by opening the Trust Center dialog box. The Trust Center will be explored in Activity 1c.

Table 8.1 Macro Security Settings for a Database Not Opened from a Trusted Location

Macro Setting	Description
Disable all macros without notification	All macros are disabled; security alerts will not appear.
Disable all macros with notification	All macros are disabled; security alert appears with the option to enable the content if the source of the file is trusted. This is the default setting.
Disable all macros except digitally signed macros	A macro that does not contain a digital signature is disabled; security alerts do not appear. If the macro is digitally signed by a publisher in the Trusted Publishers list, the macro is allowed to run. If the macro is digitally signed by a publisher not in the Trusted Publishers list, a security alert appears.
Enable all macros (not recommended; potentially dangerous code can run)	All macros are allowed; security alerts do not appear.

Activity 1c Viewing Trust Center Settings

Part 3 of 3

1. To explore current settings in the Trust Center, complete the following steps:
   a. With Access open, click the File tab and then click *Options or* click the <u>Open Other Files</u> hyperlink at the bottom of the Recent list and then click *Options*.
   b. Click *Trust Center* in the left pane of the Access Options dialog box.
   c. Click the Trust Center Settings button in the *Microsoft Access Trust Center* section.

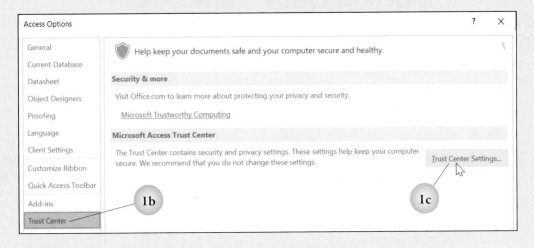

d. At the Trust Center dialog box, click *Macro Settings* in the left pane.

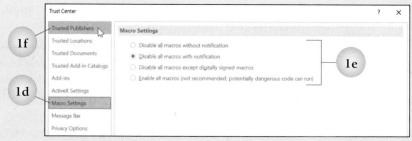

e. Review the options in the *Macro Settings* section. Notice which option is active on your computer. The default option is *Disable all macros with notification*. **Note: The security setting on your computer may be different from the default option. Do not change the security setting without the permission of your instructor.**

f. Click *Trusted Publishers* in the left pane. If any publishers have been added to the list on your computer, their names will appear in the list box. If the list box is empty, no trusted publishers have been added.

g. Click Trusted Locations in the left pane. Review the path and description of any folder in the Trusted Locations list. By default, Access adds the folder created when Microsoft Access is installed that contains the wizard database templates provided by Microsoft. Additional folders that have been added by a system administrator or network administrator may also appear.

h. Click OK to close the Trust Center dialog box.

2. Click OK to close the Access Options dialog box.

Activity 2 Import and Merge Data from External Sources   4 Parts

You will link and import data from a table in another Access database and a comma delimited text file. You will also save import specifications for an import routine that you expect to repeat often.

Tutorial

Importing Data from Another Access Database

Importing Data from Another Access Database

Data stored in another Access database can be merged with data in the active database by importing a copy of the source object. Multiple objects can be copied, including the relationships between tables. When importing, specify to import only the definition or both the definition and the data.

In Chapter 3, an append query was used to merge data from tables that contain the same fields by adding selected records to the end of an existing table. Earlier in this chapter, Activity 1 stressed the importance of backing up a database before making any major changes to it (such as merging the data) because some changes cannot be undone. If problems arise as a result of making changes to a database, either restore the entire database or import only the damaged table from the backup copy that was created.

New Data Source

To begin an import operation, click the External Data tab and then click the New Data Source button in the Import & Link group, hover over *From Database*, and then click *Access* from the drop-down list. The Get External Data - Access Database dialog box opens as shown in Figure 8.1. Specify the source database

Quick Steps

Import Objects from Access Database

1. Open destination database.
2. Click External Data tab.
3. Click New Data Source button.
4. Hover over *From Database*.
5. Click *Access*.
6. Click Browse button.
7. If necessary, navigate to drive and/or folder.
8. Double-click source file name.
9. Click OK.
10. Select import object(s).
11. Click OK.
12. Click Close.

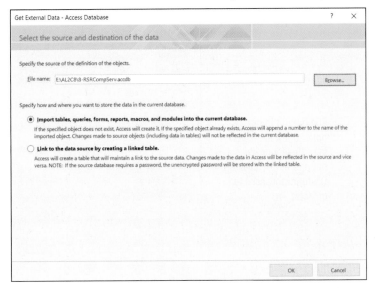

Hint When importing a query, form, or report, make sure that the tables associated with the object are also imported.

Hint If an object with the same name as an imported table already exists in the destination database, Access does not overwrite the existing object. Instead, it adds *1* to the end of the name of the imported object.

containing the object(s) to be imported by clicking the Browse button to open the File Open dialog box. Navigate to the drive and/or folder containing the source database and then double-click the Access database file name to insert the database file name in the *File name* text box. With the *Import tables, queries, forms, reports, macros, and modules into the current database* option selected by default, click OK. This opens the Import Objects dialog box, shown in Figure 8.2. Select the objects to be imported, change options if necessary, and then click OK.

Click the Options button to display the *Import*, *Import Tables*, and *Import Queries* sections, shown in Figure 8.3. By default, Access imports relationships between tables as well as their structure definitions and data. Access also imports a query as a query, as opposed to importing a query as a table. Select or deselect the options as necessary before clicking OK to begin the import operation.

An object can also be copied by opening two copies of Access: one with the source database opened and the other with the destination database opened. With the source database window active, right-click the source object in the Navigation pane and then click *Copy*. Switch to the window containing the destination database, right-click in the Navigation pane, and then click *Paste*. Close the Access window containing the source database.

Figure 8.2 Import Objects Dialog Box

Click the tab of the object type to be imported, click the object name and then click OK. Use the Shift key (adjacent objects) and the Ctrl key (nonadjacent objects) to select multiple objects.

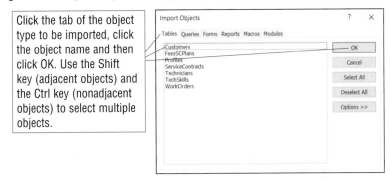

Figure 8.3 Import Objects Dialog Box with Options Displayed

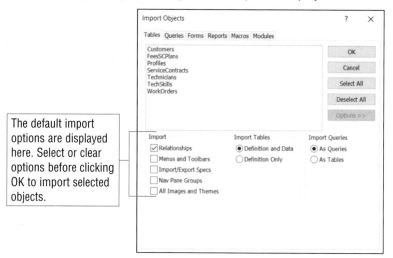

The default import options are displayed here. Select or clear options before clicking OK to import selected objects.

Activity 2a Importing a Form and Table from Another Access Database

1. Open **8-RSRTechPay** and enable the content.
2. Merge the RSRCompServe database with the RSRTechPay database by importing the WorkOrders table and TechMaintenance form from **8-RSRCompServ**. Import the two objects by completing the following steps:
   a. Click the External Data tab.
   b. Click the New Data Source button in the Import & Link group.
   c. Hover over *From Database* and then click *Access* at the drop-down list.
   d. At the Get External Data - Access Database dialog box, click the Browse button.
   e. At the File Open dialog box, double-click **8-RSRCompServ**. *Note: Navigate to the AL2C8 folder on your storage medium, if necessary*.
   f. With the *Import tables, queries, forms, reports, macros, and modules into the current database* option already selected, click OK.

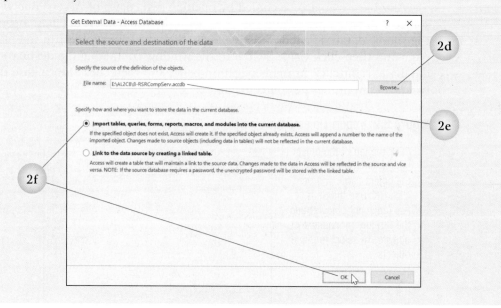

g. At the Import Objects dialog box with the Tables tab selected, click *WorkOrders* in the list box.
h. Click the Forms tab.
i. Click *TechMaintenance* in the list box and then click OK.
j. At the Get External Data - Access Database dialog box with the *Save import steps* check box empty, click the Close button.

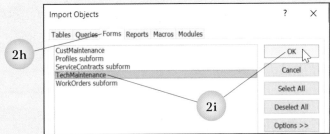

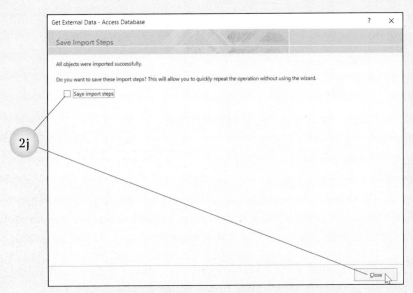

3. Access imports the WorkOrders table and TechMaintenance form and adds the object names to the Navigation pane. The form will not be operational until after Activity 2b because the tables needed to populate data in the form do not yet reside in the database. The dependent tables were not imported in this activity because the tables that contain the records are to be linked in a later activity.

Tutorial

Linking to a Table in Another Access Database

Linking to a Table in Another Access Database

In Activity 2a, a copy of a form was imported from one database to another. If the source form object is modified, the imported copy of the form object will not be altered. To ensure that the table in the destination database will reflect any changes made to the source table, link the data in the object when it is imported.

To create a linked table in the destination database, click the New Source button in the Import & Link group on the External Data tab, hover over *From Database*, and then click *Access* from the drop-down list. Click the Browse button, navigate to the drive and/or folder in which the source database is stored, and then double-click the source database file name. Click the *Link to the data source by creating a linked table* option at the Get External Data - Access Database dialog box and then click OK, as shown in Figure 8.4.

Figure 8.4 Get External Data - Access Database Dialog Box with the Link Option Selected

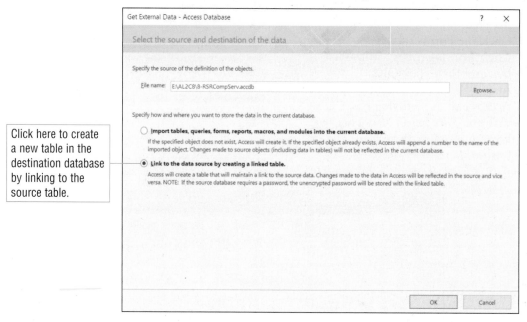

Click here to create a new table in the destination database by linking to the source table.

Quick Steps

Link to Table in Another Database
1. Open destination database.
2. Click External Data tab.
3. Click New Data Source button.
4. Hover over *From Database*.
5. Click *Access*.
6. Click Browse button.
7. If necessary, navigate to drive and/or folder.
8. Double-click source file name.
9. Click *Link to the data source by creating a linked table*.
11. Click OK.
12. Select table(s).
13. Click OK.

The Link Tables dialog box, shown in Figure 8.5, contains the list box, from which the tables to be linked are selected. Use the Shift key or the Ctrl key to select multiple tables to link in one step. Linked tables are indicated in the Navigation pane with blue right-pointing arrows.

When a table is linked, the source data does not reside in the destination database. Opening a linked table causes Access to update the datasheet with the information in the source table in the other database. Edit the source data in either the source table or the linked table in the destination database. Either way, the changes will be reflected in both tables.

Figure 8.5 Link Tables Dialog Box

Choosing Whether to Import or Link to Source Data

If the source data is not likely to change, the table should be imported. Keep in mind, however, that because importing creates copies of the data in two locations, changes or updates to the data must be made in both copies. Making the changes twice increases the risk of making a data entry error or forgetting to make an update in one or the other location.

If the source data is updated frequently, link to it so that changes have to be made only once. Because linked data exists only in the source location, the likelihood of making an error or missing an update is reduced.

Also link to the data source file when several different databases require a common table, such as an Inventory table. To duplicate the table in each database is inefficient and wastes disk space. There is also a potential for error if individual databases are not refreshed with updated data. Given these risks, it makes more sense to link rather than import. In this scenario, a master Inventory table in a separate, shared database will be linked to all the other databases that need to use the data.

Activity 2b Linking to Tables in Another Access Database

Part 2 of 4

1. With **8-RSRTechPay** open, link to two tables in **8-RSRCompServ** by completing the following steps:
   a. With the External Data tab still active, click the New Data Source button.
   b. Hover over *From Database* and then click *Access* at the drop-down list.
   c. Click the Browse button and then double-click **8-RSRCompServ** in the AL2C8 folder.
   d. Click the *Link to the data source by creating a linked table* option and then click OK.

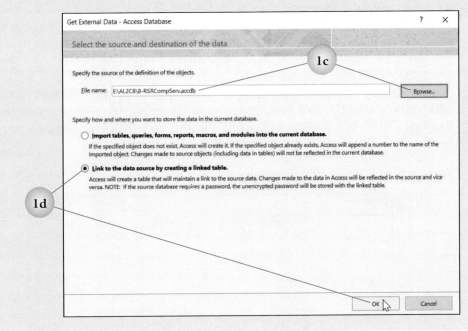

e. At the Link Tables dialog box, click *Technicians* in the list box.

f. Press and hold down the Shift key, click *TechSkills* in the list box, and then release the Shift key.

g. Click OK. Access links the two tables to the source database and adds the table names to the Navigation pane. Each linked table displays with a blue right-pointing arrow next to the table icon.

2. Double-click *TechMaintenance* in the Forms group in the Navigation pane to view the form with the first record displayed. Close the form.

3. Double-click *Technicians* in the Tables group in the Navigation pane to view the table datasheet and then close the datasheet.

4. Double-click *TechSkills* in the Tables group in the Navigation pane to view the table datasheet and then close the datasheet.

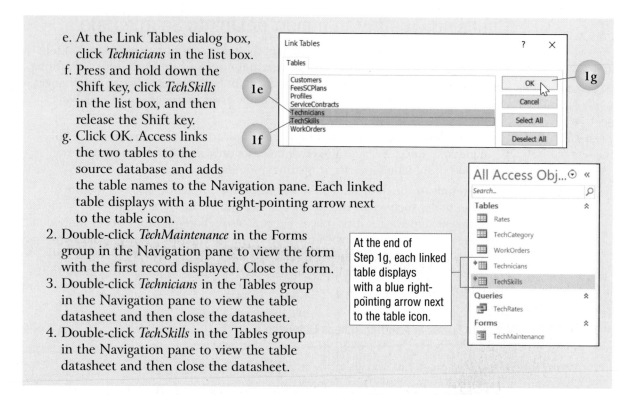

At the end of Step 1g, each linked table displays with a blue right-pointing arrow next to the table icon.

Resetting a Link **Using the Linked** **Table Manager**

Resetting a Link Using the Linked Table Manager

When a table has been linked to another database, Access stores the full path to the source database file name along with the linked table name. Changing the database file name or folder location for the source database means that the linked table will no longer function. Access provides the Linked Table Manager dialog box, shown in Figure 8.6, to allow the user to reset or refresh the link to a table to reconnect to the data source file.

To refresh a link to a table, open the Linked Table Manager dialog box by clicking the Linked Table Manager button in the Import & Link group on the External Data tab. Click the Expand All button, click the check box next to the link to be refreshed, and then click the Refresh button.

Linked Table **Manager**

Quick Steps

Refresh Link
1. Click External Data tab.
2. Click Linked Table Manager button.
3. Click Expand All button.
4. Click check box.
5. Click Refresh.
6. Click Close button.

Figure 8.6 Linked Table Manager Dialog Box

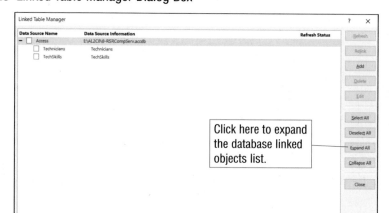

Click here to expand the database linked objects list.

1. With **8-RSRTechPay** open, change the location of **8-RSRCompServ** by completing the following steps:

   a. Click the File tab, click *Open*, click the *Browse* option in the *Recent* section, and then locate the AL2C8 folder on your storage medium.

   b. Right-click **8-RSRCompServ** in the file list box and then click *Cut* at the shortcut menu.

   c. Double-click the *DatabaseBackUps* folder.

   d. Right-click in a blank area of the file list box and then click *Paste* at the shortcut menu.

   e. Close the Open dialog box. Because the location of the source database has been moved, the linked tables are no longer connected to the correct location.

   f. Click the Back button.

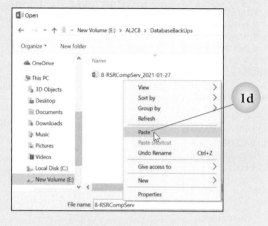

2. Refresh the links to the two tables by completing the following steps:

   a. If necessary, click the External Data tab.

   b. Click the Linked Table Manager button in the Import & Link group.

   c. At the Linked Table Manager dialog box, click the Expand All button

   d. Click the Select All button to select all the linked objects.

   e. Click the Refresh button.

   f. Because the source database has been moved, a message stating that Access could not find the file. Click OK.

   g. Failed now appears the Refresh Status column. Click the Select All button.

3. Relink the two tables by completing the following steps:

   a. Click the Relink button.

   b. At the Select New Location of Access, double-click **DatabaseBackUps** and then double-click **8-RSRCompServ**.

   c. Click Yes at the message box asking if you want to relink the selected tables in the new data source.

   d. Click OK at the Relink TechSkills dialog box to relink the TechSkills table. ***Note: You may be asked to relink the Technicians table first***.

   e. Click OK at the Relink Technicians dialog box to relink the Technicians table.

   f. With the Refresh Status changed to *Success* click the Close button.

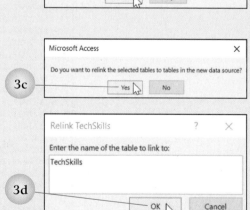

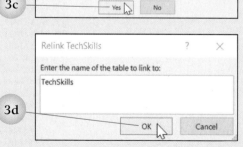

> **Check Your Work**

Importing Data into a Database from a Text File

Text files are often used to exchange data between dissimilar programs because this file format is recognized by nearly all applications. Text files contain no formatting and consist only of letters, numbers, punctuation symbols, and a few control characters. Two commonly used text file formats separate fields with a tab (delimited file format) and a comma (comma separated file format). Figure 8.7 shows a partial view of the text file used in Activity 2d in a Notepad window. If necessary, view and edit a text file in Notepad before importing if the source application inserts characters that will be deleted.

To import a text file into Access, click the External Data tab and then click the New Data Source button in the Import & Link group, hover over *From File*, and then click *Text File* from the drop-down list. Access opens the Get External Data - Text File dialog box, which is similar to the dialog box used to import data from another Access database. When importing a text file, Access provides an append option in addition to the import and link options in the *Specify how and where you want to store the data in the current database* section. Click the Browse button to navigate to the source file and double-click the source file name to launch the Import Text Wizard, which guides the user through the import process using four dialog boxes.

Figure 8.7 Activity 2d Partial View of Text File Content in Notepad

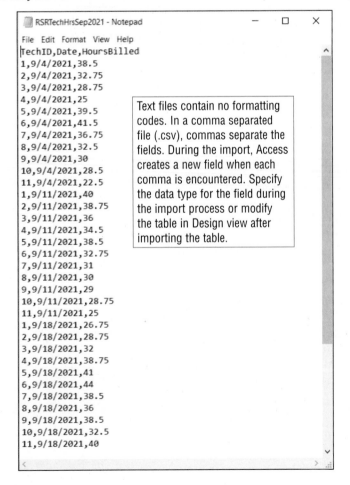

Text files contain no formatting codes. In a comma separated file (.csv), commas separate the fields. During the import, Access creates a new field when each comma is encountered. Specify the data type for the field during the import process or modify the table in Design view after importing the table.

Saving and Repeating Import Specifications

Quick Steps

Save Import Specifications
1. At last Get External Data dialog box, click *Save import steps.*
2. If necessary, edit name in *Save as* text box.
3. Type description in *Description* text box.
4. Click Save Import button.

 Saved Imports

Save import specifications for an import routine that is likely to be repeated. The last step in the Get External Data - Text File dialog box displays a *Save import steps* check box. Click the check box to expand the dialog box to display the *Save as* and *Description* text boxes as shown in Figure 8.8. Type a unique name to assign to the import routine and provide a brief description of the steps. Click the Save Import button to complete the import and store the specifications. Click the *Create Outlook Task* check box to create an Outlook task that can be set up as a recurring item for an import or export operation that is repeated at fixed intervals.

Once an import routine has been saved, repeat the import process by opening the Manage Data Tasks dialog box with the Saved Imports tab selected, as shown in Figure 8.9. To do this, click the External Data tab and then click the Saved Imports button in the Import & Link group. Click the import name and then click the Run button to instruct Access to repeat the import operation.

Figure 8.8 Get External Data - Text File Dialog Box with Save Import Steps

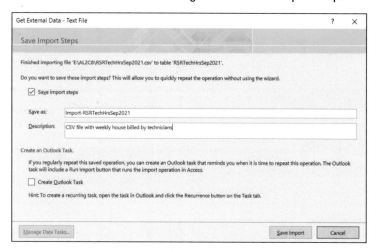

Figure 8.9 Manage Data Tasks Dialog Box with the Saved Imports Tab Selected

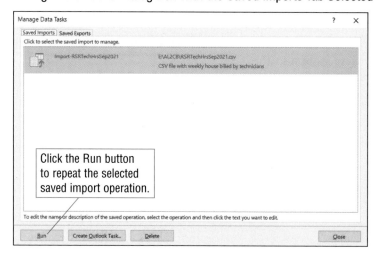

Activity 2d Importing Data from a Comma Separated Text File
and Saving Import Specifications

1. With **8-RSRTechPay** open, select a text file to import that contains the weekly hours billed for each technician for the month of September 2021 by completing the following steps:
   a. If necessary, click the External Data tab.
   b. Click the New Data Source button.
   c. Hover over *From File* and then click *Text File* at the drop-down list.
   d. At the Get External Data - Text File dialog box, click the Browse button.
   e. At the File Open dialog box, navigate to the AL2C8 folder on your storage medium if necessary.
   f. Double-click the file named *RSRTechHrsSep2021*.
   g. With *Import the source data into a new table in the current database* already selected, click OK. This launches the Import Text Wizard.

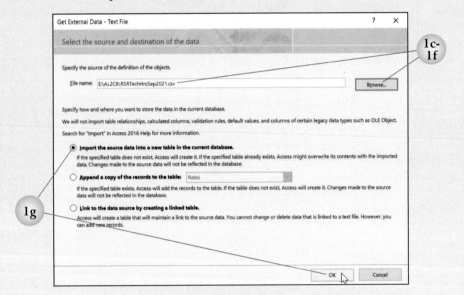

2. Import the comma separated data using the Import Text Wizard by completing the following steps:
   a. At the first Import Text Wizard dialog box with *Delimited* selected as the format, notice that the preview section in the lower half of the dialog box displays a sample of the data in the source text file. Delimited files use commas or tabs as separators while fixed width files use spaces. Click the Next button.

b. At the second Import Text Wizard dialog box with *Comma* already selected as the delimiter, click the *First Row Contains Field Names* check box. Notice that the preview section already shows the data set in columns, similar to a table datasheet. Click the Next button.

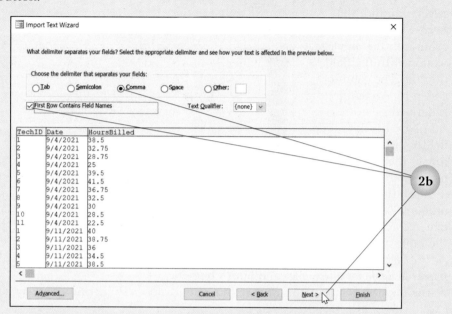

c. At the third Import Text Wizard dialog box with the *TechID* column in the preview section selected, click the *Data Type* option box arrow in the *Field Options* section and then click *Short Text* at the drop-down list.
d. Click the Next button.

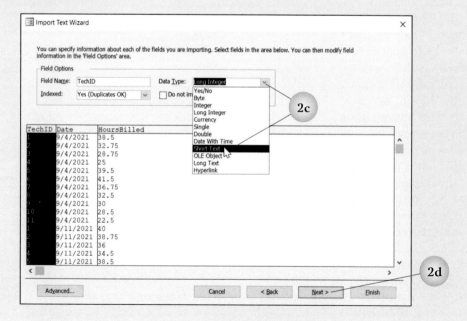

e. At the fourth Import Text Wizard dialog box with *Let Access add primary key* already selected, notice that Access has added a column in the preview section with the field title *ID*. The column added by Access is defined as an AutoNumber data type field, in which each row in the text file is numbered sequentially to make the row unique. The ID for any new record added to the table will automatically be incremented by one. Click the Next button.

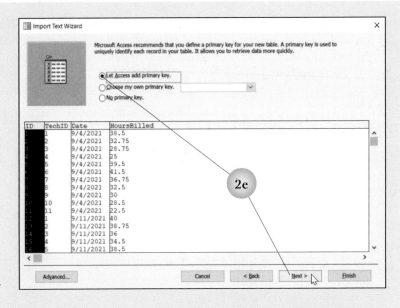

f. At the last Import Text Wizard dialog box, with *RSRTechHrsSep2021* entered in the *Import to Table* text box, click the Finish button.

3. Save the import specifications, in case this import needs to be run again, by completing the following steps:

a. At the Get External Data - Text File dialog box, click the *Save import steps* check box. This causes the *Save as* and *Description* text boxes to appear, as well as the *Create an Outlook Task* section. By default, Access creates a name in the *Save as* text box and *Import* precedes the file name containing the imported data.

b. Click in the *Description* text box and then type CSV file with weekly hours billed by technicians.

c. Click the Saved Import button.

4. Double-click *RSRTechHoursSep2021* in the Tables group in the Navigation pane to open the table in Datasheet view.

5. Print the datasheet with the bottom margin set to 0.5 inch and then close the datasheet.

6. Close **8-RSRTechPay**.

Check Your Work

Activity 3 Export Access Data to a Text File 3 Parts

You will export a query as a comma delimited text file and another query as a tab delimited text file, including saving the second export steps so you can repeat the export operation.

Tutorial

Exporting Access
Data to a Text File

Exporting Access Data to a Text File

The Export group on the External Data tab contains buttons to export Access data from a table, query, form, or report to other applications, such as Excel and Word. To work with data from other programs, click the More button in the Export group to see if a file format converter exists for the application to be used. For example, the More button contains options to export in Word, SharePoint List, ODBC Database, and HTML Document format.

Export to text file

If a file format converter does not exist for the program to be used, export the data as a text file; most applications recognize and can import a text data file. Access provides the Export Text Wizard, which is launched after an object is selected in the Navigation pane. Click the Export to text file button (which displays the text *Text File*) in the Export group on the External Data tab and then specify the name of the exported text file and where to store it. The steps in the Export Text Wizard are similar to those in the Import Text Wizard, used in Activity 2d.

Activity 3a Exporting a Query as a Text File Part 1 of 3

1. Display the Open dialog box and then use the Cut and Paste features to move **8-RSRCompServ** back from the DatabaseBackUps folder to the AL2C8 folder on your storage medium.
2. Open **8-RSRCompServ** and enable the content if necessary.
3. Export the *TotalWorkOrders* query as a text file by completing the following steps:
    a. Select the TotalWorkOrders query in the Navigation pane.
    b. Click the External Data tab.
    c. Click the Export to text file button (which displays the text *Text File*) in the Export group.
    d. At the Export - Text File dialog box, click the Browse button.
    e. At the File Save dialog box, navigate to the AL2C8 folder on your storage medium if necessary.
    f. With the default file name *TotalWorkOrders.txt* in the *File name* text box, click the Save button.
    g. Click OK.

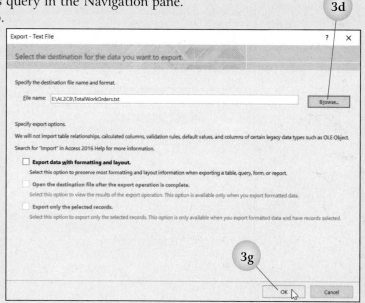

h. At the first Export Text Wizard dialog box with *Delimited* selected as the format, notice in the preview section of the dialog box that commas separate the fields and that data in a field defined with the Text data type is enclosed in quotation marks. Click the Next button.

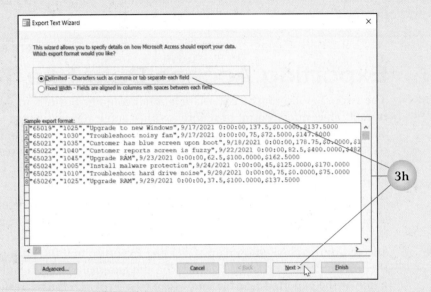

i. At the second Export Text Wizard dialog box with *Comma* selected as the delimiter character that separates the fields, click the *Include Field Names on First Row* check box to insert a check mark. Note that Access adds a row containing the field names to the top of the data in the preview section. Each field name is enclosed in quotation marks.

j. Click the arrow next to the *Text Qualifier* option box and then click {*none*} at the drop-down list. Access removes all the quotation marks from the text data in the preview section.

k. Click the Next button.

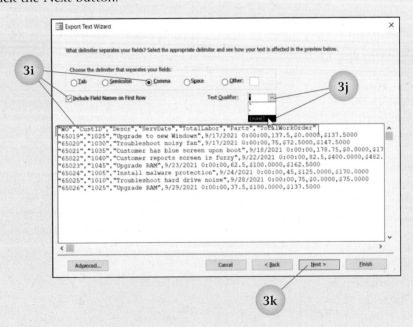

l. At the last Export Text Wizard dialog box, with *[E]:\AL2C8\TotalWorkOrders.txt* (where *[E]* is the drive for your storage medium) entered in the *Export to File* text box, click the Finish button.

m. Click the Close button to close the dialog box without saving the export steps.

4. Open Notepad and view the text file by completing the following steps:

a. Click the Start button, and then start typing notepad. When *Notepad* displays in the *Best match* section, press the Enter key. (Depending on your operating system, these steps may vary).

b. At a blank Notepad window, click the File button and then click *Open*. Navigate to the AL2C8 folder on your storage medium and then double-click **TotalWorkOrders.**

5. Click the File button and then click *Print* to print the exported text file.

6. Exit Notepad.

Saving and Repeating a Saved Import or Export Specifications

Saving and Repeating Export Specifications

Just as import specifications can be saved for reuse, so can export specifications. The last Export - Text File dialog box displays a *Save export steps* check box. Click the check box to expand the dialog box options and display the *Save as* and *Description* text boxes. Type a unique name for the export routine and a brief description of the steps. Click the Save Export button to complete the export operation and store the specifications for later use.

Quick Steps

Save Export Specifications
1. At last Export - Text File dialog box, click *Save export steps* check box.
2. Verify file name.
3. Type description in *Description* text box.
4. Click Save Export button.

Activity 3b Exporting a Query as a Text File and Saving the Export Steps   Part 2 of 3

1. With **8-RSRCompServ** open, double-click the *TotalLabor* query in the Navigation pane. Notice that only the work orders for the month of September are retrieved. Switch to Design view and see that *DatePart("m",[ServDate])* has been added right of the *TotalLabor* field and that *9* displays in the *Criteria* row. With this criteria, the query returns only the results for September, the ninth month. Close the query.

2. Export the TotalLabor query as a text file using tabs as the delimiter characters by completing the following steps:

a. With the TotalLabor query selected in the Navigation pane, click the External Data tab and then click the Text File button in the Export group.

b. With *[E]:\AL2C8\TotalLabor.txt* (where *[E]* is the drive for your storage medium) entered in the *File name* text box, click OK.

c. Complete the steps of the Export Text Wizard as follows:
1) With *Delimited* selected in the first dialog box, click the Next button.
2) At the second dialog box, choose *Tab* as the delimiter character, click the *Include Field Names on First Row* check box to insert a check mark, change the *Text Qualifier* option to *{none}*, and then click the Next button.

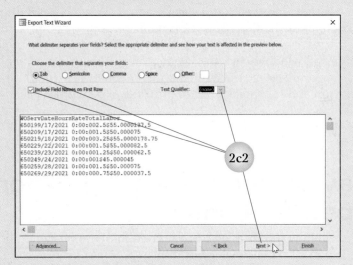

3) Click the Finish button.
d. At the Export - Text File dialog box, click the *Save export steps* check box to insert a check mark.
e. Click in the *Description* text box and then type TotalLabor query for RSR Computer Service work orders as a text file.
f. Click the Save Export button.

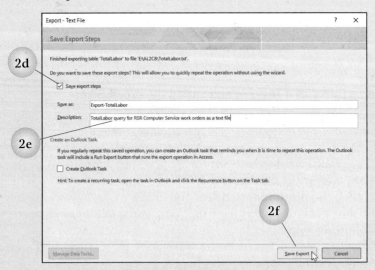

3. Start Notepad.
4. At a blank Notepad window, open **TotalLabor**.
5. Print **TotalLabor** and then exit Notepad.

 Check Your Work

 Saved Exports

Quick Steps

Repeat Export Steps
1. Click External tab.
2. Click Saved Exports button.
3. Click specified saved export.
4. Click Run button.
5. Click OK.

Once an export routine has been saved, repeat the export process by clicking the Saved Exports button in the Export group on the External Data tab. This opens the Manage Data Tasks dialog box with the Saved Exports tab selected, as shown in Figure 8.10. Click the export name in the dialog box and then click the Run button to instruct Access to repeat the export operation. Be careful not to override any previous exported files because the file name and location will be the same.

Figure 8.10 Manage Data Tasks Dialog Box with the Saved Exports Tab Selected

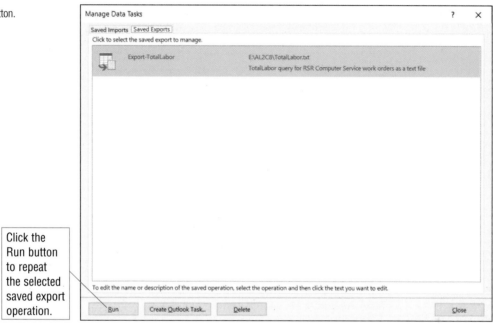

Click the Run button to repeat the selected saved export operation.

Activity 3c Repeating Export Steps

Part 3 of 3

1. With **8-RSRCompServ** open, change the location of **TotalLabor** text file by completing the following steps:
   a. Click the File tab, click *Open*, click the Browse button, and then locate the AL2C8 folder on your storage medium.
   b. Click the *Data Type* option box that currently displays *Microsoft Access* and change the option to *All Files*.
   c. Scroll down, right-click *TotalLabor* in the file list box, and then click *Cut* at the shortcut menu.
   d. Double-click the *DatabaseBackUps* folder.
   e. Right-click in a blank area of the file list box and then click *Paste*.
   f. Close the Open dialog box. The Export procedure can now be run without overwriting the data for the month of September. Click the Back button.

2. Export the results of the TotalLabor query for October by completing the following steps:
   a. Open the TotalLabor query in Design view.
   b. In the *Criteria* row of the *DatePart("m",[ServDate])* field, change the *9* to a *10*. Changing this number will return the data for the month of October.
   c. Save, run, and then close the query.
   d. Click the External Data tab.
   e. Click the Saved Exports button in the Export group.
   f. With *Export-TotalLabor* selected, click the Run button.
   g. Click OK.
   h. Click the Close button.

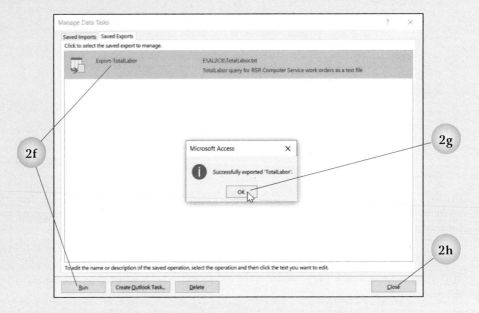

3. Click the Close button in the upper right corner to close Access.

Check Your Work

Chapter Summary

- Use options at the Save As backstage area to create backup database files on a regular basis. This will help to minimize any data loss due to system failures or design mistakes.

- Create a backup copy of a database before making any major changes to it.

- An ACCDE file is a locked-down version of a database, in which users do not have access to Design view or Layout view and therefore cannot make changes to the design of objects.

- Create an ACCDE database file at the Save As backstage area.

- Open the Access Options dialog box, click *Trust Center* in the left pane, and then click the Trust Center Settings button to view and/or modify Trust Center options.

- An object in another Access database can be imported into the active database using the Import Access database button in the Import & Link group on the External Data tab.

- If the source object is a table, choose to import or link to the source table.

- When a table is linked, the data is not copied into the active database; it resides only in the source database.
- Edit source data in a linked table in the source database or the destination database; the changes will be reflected in both tables.
- Import the table if the source data is not likely to change.
- Link to a data source file that is updated frequently so changes have to be made only once; this reduces the potential for making data entry mistakes or missing updates.
- Link to a source table that is shared among several different databases within an organization.
- When a table has been linked to another database, Access stores the full path to the source database. If the location of the source database is moved or the file name is changed, the links will need to be refreshed.
- A text file is often used to exchange data between dissimilar programs because this file format is recognized by nearly all applications.
- Import a text file into an Access database by clicking the Import text file button in the Import & Link group on the External Data tab.
- When a text file is selected for import, Access launches the Text Import Wizard, which guides the user through the steps to import the text into a table.
- If an import routine is repeated often, consider saving the steps so the routine can be run without having to perform each step every time.
- Open the Manage Data Tasks dialog box to run a saved import by clicking the Saved Imports button in the Import & Link group on the External Data tab.
- Export Access data in a text file format using the Export Text Wizard by clicking the Export to text file button in the Export group on the External Data tab.
- Within the Export Text Wizard, the user is prompted to choose the text format, delimiter character, field names, text qualifier symbols, export path, and file name.
- To repeat an export routine, first save the export steps at the last Export - Text File dialog box.
- Click the Saved Exports button in the Export group on the External Data tab to run a saved export routine.

Commands Review

FEATURE	RIBBON TAB, GROUP	BUTTON
create ACCDE file	File, *Save As*	
export data as text file	External Data, Export	
import data from text file	External Data, Import & Link	
import or link data from Access database	External Data, Import & Link	
Linked Table Manager	External Data, Import & Link	
saved exports	External Data, Export	
saved imports	External Data, Import & Link	

Index

moving to another section in custom report, 134–135

Large Number data type, 7

left outer join, 59, 60, 62

Line button, 108

Linked Table Manager button, 256

Linked Table Manager dialog box, 256–257

linking
  choosing to import or link to source data, 255
  resetting link using Linked Table Manager, 256–257
  table to another Access database, 253–257

Link Tables dialog box, 254

list box, adding
  to form, 116–118
  to report, 163–167

List Box button, 116, 163

List Box Wizard, 116

logical functions, 70

logo, adding
  to form, 108–110
  to report, 144–146

Logo button, 108

Long Text data type, 7
  Append Only property, 21
  commonly used format codes for, 12
  custom formatting for, 12
  enabling rich text formatting, 21–23
  maintaining history of changes in, 21–23

lookup field
  assigning multiple values to lookup list, 46
  create relationship with Applications Parts template, 178–179
  data from another table, 40–42
  modifying lookup list properties, 43

with multiple-value field, 44–46

Lookup Wizard
  field that stores multiple values, 44–46
  Look Up values in another table, 40–42

Lookup Wizard data type, 7

M

Macro Builder window, 208, 213

Macro button, 208, 209

macros, 208–222
  actions, 208
  arguments, 209
  command button to run, 215–219
  converting to visual basic, 221–222
  creating
    by Action Catalog task pane, 212–213
    by dragging and dropping object, 212
    to open form and find record, 210–211
    overview of, 208–209
  defined, 208
  deleting, 213–215
  editing, 213–215
  in Navigation pane, 208
  security settings for, 248–249
  viewing embedded macros, 219–222

Macro Tools Design tab, 221

main form, 96

main report, 136

Make Table button, 74, 75

Make Table dialog box, 75

make-table query, 74, 75–76

Manage Data Tasks dialog box, 259

many-to-many relationships
  establishing, 36–37
  junction table, 37–38

Maximum function, 150

menus, limiting, in database, 228

merging data, from another Access database, 250–253

Minimum function, 150

More Forms button, 113

More Options button, 150, 156

move handle, 90

moving, control objects, 90–92

multiple-field primary key, 39–40

multiple-value fields, using in queries, 72–73

N

Name & Caption dialog box, 9

Navigation button, 223

Navigation form
  adding command button to, 225–226
  creating, 222–226
  editing, 225–226

Navigation Options dialog box, 229–230

Navigation pane
  customizing, 228–230
  deleting macro from, 214
  macros in, 208
  show or hide on startup, 226–227, 228
  subform object in, 97
  subreport object in, 136–137

New backstage area, available database templates, 172–173

New Data Source button, 250

New Tab button, 233

normalizing database, 49

Normal mode, 209

Notepad, 258

Number data type, 7
  commonly used format codes for, 14
  custom format for, 14–16

O

objects
 adjusting for consistency of appearance, 105–108
 creating using application parts template, 177–184
 deleting, 200–201
 dragging and dropping object to create macro, 212
 optimizing performance using Performance Analyzer, 192–195
 renaming, 200–201
 subform object, 97
 subreport object, 136–137
 tab control, 95–99
OLE Object data type, 7
one-to-many relationships
 editing options, 33–35
 establishing, 30–33
 foreign key, 30
 orphan record, 34
one-to-one relationship, establishing, 36–37
Open backstage area, 172
OpenForm action, 209, 212
OpenQuery action, 212
OpenReport action, 212
OpenTable action, 212
operator, in SQL, 202
orphan records, 34
outer join, 59–60, 62

P

page break, inserting before and after sections, 156
Page Footer section, 131
Page Header section, 131
Page Number dialog box, 140–141
page numbers, adding to reports, 140–144
Page Numbers button, 140
parameter query
 defined, 56

to prompt for starting/ending date, 58
to prompt for technician names, 56–57
uses of, 56
Pass-Through query, 74
Paste button, 186
Paste Table dialog box, 186–187
Performance Analyzer, 192–195
Performance Analyzer dialog box, 192–193
primary key
 composite key, 39
 defining multiple-field primary key, 39–40
 in one-to-many relationship, 30
Primary Key button, 39
primary key field, 5
 diagramming databases, 6
 set as Required property, 10
printing, relationships, 36
Priority buttons, 147
Property Sheet task pane, 64, 87, 96, 155–156

Q

queries, 53–80
 action queries, 74–80
  adding records to table using, 78–79
  creating new table with, 75–76
  deleting group of records, 77
  modifying records using, 79–80
 adding/removing tables from, 62–64
 alias for table, 64
 append query, 74, 78–79
 Cartesian product query, 66
 conditional logic in, 70–71
 creating using Application Parts template, 179–181
 cross product query, 66
 delete query, 74, 77

exporting query as text file, 263–265
join properties in, 59–66
 inner join, 59, 61
 left outer join, 59, 60, 62
 outer join, 59–60
 right outer join, 59, 60, 63–64
 self-join query, 64–66
 specifying join type, 59–62
make-table query, 74, 75–76
modifying using SQL, 202–204
multiple-value fields in, 72–73
parameter query, 56–58
prompting for criteria using parameter query, 56–57
running with no established relationship, 66
saving a filter as, 54–55
select query, 54–58
subqueries, 67–69
TotalLabor query, 203–204
update query, 74, 79–80
Query Design button, 56
Query Design view, 54
Query Tools Design tab, 62, 74
question mark, as wildcard character with Find feature, 121
Quick Access Toolbar
 customizing, 238–239
 resetting, 240
Quick Start options, 177–178

R

Read Only mode, 209
records
 adding, using query, 78–79
 attaching files to, 23–25
 deleting, using query, 77
 finding and sorting in forms, 120–122
 grouping in report, 146–149
 macro to open form and find record, 210–211

Interior Photo Credits

Page GS-1 (banner image) © lowball-jack/GettyImages; *page GS-1, (in Figure G.1)* all images courtesy of Paradigm Education Solutions; *page GS-2,* © irbis picture/Shutterstock.com; *page GS-3,* © th3fisa/Shutterstock.com; *page GS-4,* © goldyg/Shutterstock.com; *page GS-5,* © Pressmaster/Shutterstock.com.